Environmental Science, Engineering and Technology

# Paleoecological Significance of Diatoms in Argentinean Estuaries

# ENVIRONMENTAL SCIENCE, ENGINEERING AND TECHNOLOGY

**Nitrous Oxide Emissions Research Progress**
*Adam I. Sheldon and Edward P. Barnhart (Editors)*
2009. ISBN: 978-1-60692-267-5

**Fundamentals and Applications of Biosorption Isotherms, Kinetics and Thermodynamics**
*Yu Liu and Jianlong Wang (Editors)*
2009. ISBN: 978-1-60741-169-7

**Environmental Effects of Off-Highway Vehicles**
*Douglas S. Ouren, Christopher Haas, Cynthia P. Melcher, Susan C. Stewart, Phadrea D. Ponds, Natalie R. Sexton, Lucy Burris, Tammy Fancher and Zachary H. Bowen*
2009. ISBN: 978-1-60692-936-0

**Agricultural Runoff, Coastal Engineering and Flooding**
*Christopher A. Hudspeth and Timothy E. Reeve (Editors)*
2009. ISBN: 978-1-60741-097-3
2009. ISBN: 978-1-60876-608-6 (E-book)

**Soil Fertility**
*Derek P. Lucero and Joseph E. Boggs (Editors)*
2009. ISBN: 978-1-60741-466-7

**Conservation of Natural Resources**
*Nikolas J. Kudrow (Editor)*
2009. ISBN: 978-1-60741-178-9
2009. ISBN: 978-1-60876-642-6 (E-book)

**Directory of Conservation Funding Sources for Developing Countries: Conservation Biology, Education and Training, Fellowships and Scholarships**
*Alfred O. Owino and Joseph O. Oyugi*
2009. ISBN: 978-1-60741-367-7

**Forest Canopies: Forest Production, Ecosystem Health and Climate Conditions**
*Jason D. Creighton and Paul J. Roney (Editors)*
2009. ISBN: 978-1-60741-457-5

**Handbook of Environmental Policy**
*Johannes Meijer and Arjan der Berg (Editors)*
2009. ISBN: 978-1-60741-635-7

**Handbook on Environmental Quality**
*Evan K. Drury and Tylor S. Pridgen (Editors)*
2009. ISBN: 978-1-60741-420-9
2009. ISBN: 978-1-61728-018-4 (E-book)

**Handbook of Environmental Research**
*Aurel Edelstein and Dagmar Bär (Editors)*
2009. ISBN: 978-1-60741-492-6

**Environmental Cost Management**
*Randi Taylor Mancuso*
2009. ISBN: 978-1-60741-815-3

**The Amazon Gold Rush and Environmental Mercury Contamination**
*Daniel Marcos Bonotto and Ene Glória da Silveira*
2009. ISBN: 978-1-60741-609-8

**Biological and Environmental Applications of Gas Discharge Plasmas**
*Graciela Brelles-Mariño (Editor)*
2009. ISBN: 978-1-60741-945-7

**Buildings and the Environment**
*Jonas Nemecek and Patrik Schulz (Editors)*
2009. ISBN: 978-1-60876-128-9

**River Sediments**
*Greig Ramsey and Seoras McHugh (Editors)*
2009. ISBN: 978-1-60741-437-7

**Tree Growth: Influences, Layers and Types**
*Wesley P. Karam (Editor)*
2009. ISBN: 978-1-60741-784-2

**Environmental Modeling with GPS**
*Lubos Matejicek (Editor)*
2010. ISBN: 978-1-60876-363-4

**Sorbents: Properties, Materials and Applications**
*Thomas P. Willis (Editor)*
2009. ISBN: 978-1-60741-851-1
2009. ISBN: 978-1-61668-308-5 (E-book)

**Syngas: Production Methods, Post Treatment and Economics**
*Adorjan Kurucz and Izsak Bencik (Editors)*
2009. ISBN: 978-1-60741-841-2
2009. ISBN: 978-1-61668-214-9 (E-book)

**Process Engineering in Plant-Based Products**
*Hongzhang Chen*
2009. ISBN: 978-1-60741-962-4

**Potential of Activated Sludge Utilization**
*Xiaoyi Yang*
2009. ISBN: 978-1-60876-019-0

**Recent Progress on Earthquake Geology**
*Pierpaolo Guarnieri (Editor)*
2009. ISBN: 978-1-60876-147-0

**Fluid Waste Disposal**
*Kay W. Canton (Editor)*
2010. ISBN: 978-1-60741-915-0

**Estimating Future Recreational Demand**
Peter T. Yao (Editor)
2009. ISBN: 978-1-60692-472-3

**Bioengineering for Pollution Prevention**
Dianne Ahmann and John R. Dorgan
2009. ISBN: 978-1-60692-900-1
2009. ISBN: 978-1-60876-574-4
(E-book)

**Aquifers: Formation, Transport and Pollution**
Rachel H. Laughton (Editor)
2010. ISBN: 978-1-61668-051-0
2010. ISBN: 978-1-61668-444-0
(E-book)

**Temporarily Open/Closed Estuaries in South Africa**
R. Perissinotto, D.D. Stretch, A.K. Whitfield, J.B. Adams, A.T. Forbes and N.T. Demetriades
2010. ISBN: 978-1-61668-412-9
2010. ISBN: 978-1-61668-825-7
(E-book)

**Carbon Capture and Storage including Coal-Fired Power Plants**
Todd P. Carington (Editor)
2010. ISBN: 978-1-60741-196-3

**Handbook on Agroforestry: Management Practices and Environmental Impact**
Lawrence R. Kellimore (Editor)
2010. ISBN: 978-1-60876-359-7

**Biodiversity Hotspots**
Vittore Rescigno and Savario Maletta (Editors)
2010. ISBN: 978-1-60876-458-7

**Mine Drainage and Related Problems**
Brock C. Robinson (Editor)
2010. ISBN: 978-1-60741-285-4
2010. ISBN: 978-1-61668-643-7
(E-book)

**Pipelines for Carbon Sequestration: Background and Issues**
Elvira S. Hoffmann (Editor)
2010. ISBN: 978-1-60741-383-7

**The Role of Forests in Carbon Capture and Climate Change**
Roland Carnell (Editor)
2010. ISBN: 978-1-60741-447-6

**Clean Fuels in the Marine Sector**
Environmental Protection Agency
2010. ISBN: 978-1-60741-275-5
2010. ISBN: 978-1-61668-431-0
(E-book)

**Freshwater Ecosystems and Aquaculture Research**
Felice De Carlo and Alessio Bassano (Editors)
2010. ISBN: 978-1-60741-707-1

**Aviation and Climate Change**
George T. Blumenthal (Editor)
2010. ISBN: 978-1-60876-757-1

**Species Diversity and Extinction**
*Geraldine H. Tepper (Editor)*
2010. ISBN: 978-1-61668-343-6
2010. ISBN: 978-1-61668-406-8
(E-book)

**Harmful Algal Blooms – Impact and Response**
*Vladimir Buteyko*
2010. ISBN: 978-1-60741-665-4

**Estuaries: Types, Movement Patterns and Climatical Impacts**
*Julian R. Crane and Ashton E. Solomon (Editors)*
2010. ISBN: 978-1-60876-859-2

**Built Environment: Design, Management and Applications**
*Paul S. Geller (Editor)*
2010. ISBN: 978-1-60876-915-5

**Wildfires and Wildfire Management**
*Kian V. Medina (Editor)*
2010. ISBN: 978-1-60876-009-1

**HYDRO GIS: Theory and Lessons from the Vietnamese Delta**
*Shigeko Haruyama and Le Thie Viet Hoa*
2010. ISBN: 978-1-60876-156-2

**Anthropology of Mining in Papua New Guinea Greenfields**
*Benedict Young Imbun*
2010. ISBN: 978-1-61668-485-3

**Grassland Biodiversity: Habitat Types, Ecological Processes and Environmental Impacts**
*Johan Runas and Theodor Dahlgren (Editors)*
2010. ISBN: 978-1-60876-542-3

**Check Dams, Morphological Adjustments and Erosion Control in Torrential Streams**
*Carmelo Consesa Garcia and Mario Aristide Lenzi (Editors)*
2010. ISBN: 978-1-60876-146-3

**Psychological Approaches to Sustainability: Current Trends in Theory, Research and Applications**
*Victor Corral-Verdugo, Cirilo H. Garcia-Cadena and Martha Frias-Armenta (Editors)*
2010. ISBN: 978-1-60876-356-6

**Pollen: Structure, Types and Effects**
*Benjamin J. Kaiser (Editor)*
2010. ISBN: 978-1-61668-669-7
2010. ISBN: 978-1-61728-048-1
(E-book)

**Mechanisms of Cadmium Toxicity to Various Trophic Saltwater Organisms**
*Zaosheng Wang, Changzhou Yan, Hainan Kong and Deyi Wu*
2010. ISBN: 978-1-60876-646-8

**Wood: Types, Properties, and Uses**
*Lorenzo F. Botannini (Editor)*
2010. ISBN: 978-1-61668-837-0
2010. ISBN: 978-1-61728-046-7
(E-book)

**Eco-City and Green Community: The Evolution of Planning Theory and Practice**
*Zhenghong Tang (Editor)*
2010. ISBN: 978-1-60876-811-0

**Paleoecological Significance of Diatoms in Argentinean Estuaries**
*Gabriela S. Hassan*
2010. ISBN: 978-1-60876-953-7

**Geomatic Solutions for Coastal Environments**
*M. Maanan and M. Robin (Editors)*
2010. ISBN: 978-1-61668-140-1

**Marine Research and Conservation in the Coral Triangle: The Wakatobi National Park**
*Julian Clifton, Richard K.F. Unsworth and David J. Smith (Editors)*
2010. ISBN: 978-1-61668-473-0

**Modelling Flows in Environmental and Civil Engineering**
*F. Kerger, B.J. Dewals, S. Erpicum, P. Archambeau and M. Pirotton*
2010. ISBN: 978-1-61668-662-8
2010. ISBN: 978-1-61668-490-7
(E-book)

**Natural Resources in Ghana: Management, Policy and Economics**
*David M. Nanang and Thompson K. Nunifu (Editors)*
2010. ISBN: 978-1-61668-020-6

**Fundamentals of General Ecology, Life Safety and Environment Protection**
*Mark D. Goldfein, Alexei V. Ivanov and Nikolaj V. Kozhevnikov*
2010. ISBN: 978-1-61668-176-0
2010. ISBN: 978-1-61668-397-9
(E-book)

**Zinc, Copper, or Magnesium Supplementation Against Cadmium Toxicity**
*Vesna Matović, Zorica Plamenac Bulat, Danijela Đukić-Ćosić and Danilo Soldatović*
2010. ISBN: 978-1-61668-332-0
2010. ISBN: 978-1-61668-721-2
(E-book)

**Sources and Reduction of Greenhouse Gas Emissions**
*Steffen D. Saldana (Editor)*
2010. ISBN: 978-1-61668-856-1
2010. ISBN: 978-1-61728-091-7
(E-book)

**Biogeography**
*Mihails Gailis and Stefans Kalniòð (Editors)*
2010. ISBN: 978-1-60741-494-0

**International Trade and Environmental Justice: Toward a Global Political Ecology**
*Alf Hornborg and Andrew K. Jorgenson (Editors)*
2010. ISBN: 978-1-60876-426-6

**Amazon Basin: Plant Life, Wildlife and Environment**
*Nicolás Rojas and Rafael Prieto (Editors)*
2010. ISBN: 978-1-60741-463-6

**Behavioral and Chemical Ecology**
*Wen Zhang and Hong Liu (Editors)*
2010. ISBN: 978-1-60741-099-7

**Global Environmental Policies: Impact, Management and Effects**
*Riccardo Cancilla and Monte Gargano (Editors)*
2010. ISBN: 978-1-60876-204-0

**Tundras: Vegetation, Wildlife and Climate Trends**
*Beltran Gutierrez and Cristos Pena (Editors)*
2010. ISBN: 978-1-60876-588-1

**Advanced Biologically Active Polyfunctional Compounds and Composites: Health, Cultural Heritage and Environmental Protection**
*Nodar Lekishvili, Gennady Zaikov and Bob Howell (Editors)*
2010. ISBN: 978-1-60876-114-2

**How Globalization is Changing the U.S. Forest Sector**
*Peter Ince, Albert Schuler, Henry Spelter and William Luppold (Editors)*
2010. ISBN: 978-1-60876-132-6

**Environmental Modeling with GPS**
*Lubos Matejicek (Editor)*
2010. ISBN: 978-1-60876-363-4

**Watersheds: Management, Restoration and Environmental Impact**
*Jeremy C. Vaughn (Editor)*
2010. ISBN: 978-1-61668-667-3
2010. ISBN: 978-1-61728-243-0 (E-book)

**A True Tale of Science and Discovery**
*Lawrence A. Curtis*
2010. ISBN: 978-1-60876-595-9

**Advances in Environmental Modeling and Measurements**
*Dragutin T. Mihailovic and Branislava Lalic (Editors)*
2010. ISBN: 978-1-60876-599-7

**Protecting the Great Lakes from Invasive and Nonindigenous Species**
*Clara E. Wouters (Editor)*
2010. ISBN: 978-1-61728-103-7
2010. ISBN: 978-1-61728-330-7 (E-book)

**Alaskan Native Villages Threatened by Erosion**
*Russell M. Trevino (Editor)*
2010. ISBN: 978-1-60876-890-5

**Wildfires, Fuels and Invasive Plants**
*Louise E. Willems (Editor)*
2010. ISBN: 978-1-61728-164-8
2010. ISBN: 978-1-61728-322-2 (E-book)

**Spatial Assemblages of Tropical Intertidal Rocky Shore Communities in Ghana, West Africa**
*Emmanuel Lamptey, Ayaa Kojo Armah and Lloyd Cyril Allotey*
2010. ISBN: 978-1-61668-767-0
2010. ISBN: 978-1-61728-448-9 (E-book)

ENVIRONMENTAL SCIENCE, ENGINEERING AND TECHNOLOGY

# PALEOECOLOGICAL SIGNIFICANCE OF DIATOMS IN ARGENTINEAN ESTUARIES

## GABRIELA S. HASSAN

*Instituto de Geología de Costas y del Cuaternario*
*Universidad Nacional de Mar del Plata*
*Consejo Nacional de Investigaciones Científicas y Técnicas (CONICET)*

Nova Science Publishers, Inc.
*New York*

Copyright © 2010 by Nova Science Publishers, Inc.

**All rights reserved.** No part of this book may be reproduced, stored in a retrieval system or transmitted in any form or by any means: electronic, electrostatic, magnetic, tape, mechanical photocopying, recording or otherwise without the written permission of the Publisher.

For permission to use material from this book please contact us:
Telephone 631-231-7269; Fax 631-231-8175
Web Site: http://www.novapublishers.com

### NOTICE TO THE READER

The Publisher has taken reasonable care in the preparation of this book, but makes no expressed or implied warranty of any kind and assumes no responsibility for any errors or omissions. No liability is assumed for incidental or consequential damages in connection with or arising out of information contained in this book. The Publisher shall not be liable for any special, consequential, or exemplary damages resulting, in whole or in part, from the readers' use of, or reliance upon, this material.

Independent verification should be sought for any data, advice or recommendations contained in this book. In addition, no responsibility is assumed by the publisher for any injury and/or damage to persons or property arising from any methods, products, instructions, ideas or otherwise contained in this publication.

This publication is designed to provide accurate and authoritative information with regard to the subject matter covered herein. It is sold with the clear understanding that the Publisher is not engaged in rendering legal or any other professional services. If legal or any other expert assistance is required, the services of a competent person should be sought. FROM A DECLARATION OF PARTICIPANTS JOINTLY ADOPTED BY A COMMITTEE OF THE AMERICAN BAR ASSOCIATION AND A COMMITTEE OF PUBLISHERS.

LIBRARY OF CONGRESS CATALOGING-IN-PUBLICATION DATA

Paleoecological signifance of diatoms in Argentinean estuaries / Gabriela S. Hassan.
  p. cm.
 Includes index.
 ISBN 978-1-60876-953-7 (softcover)
 1. Diatoms, Fossil--Argentina. 2. Estuaries--Argentina. 3. Paleoecology--Argentina. I. Title.
 QE955.H37 2009
 561'.9350982--dc22
                    2009054175

*Published by Nova Science Publishers, Inc. † New York*

# CONTENTS

| | | |
|---|---|---:|
| **Preface** | | xiii |
| **Abstract** | | xv |
| **Introduction** | | xvii |
| **Chapter 1** | Do Estuarine Diatoms Reliably Reflect Estuarine Environmental Conditions? | 1 |
| **Chapter 2** | How Much Information About Ecological Requirements of Estuarine Diatoms Do We Have? | 5 |
| **Chapter 3** | How Can Researchers Improve the Quality of Diatom-Based Paleoenvironmental Inferences in Coastal Settings of Argentina? | 33 |
| **Chapter 4** | Conclusions | 41 |
| **Acknowledgments** | | 45 |
| **References** | | 47 |
| **Appendix I.** | | 63 |
| **Appendix II.** | | 65 |
| **Index** | | 104 |

# PREFACE

In this book, the literature on modern estuarine diatoms from Argentina is revised in order to synthesize the available ecological information and to detect possible modern analogues for Quaternary diatom assemblages. The main objective is to build bridges between ecology and paleoecology, and to discuss the reaches and limitations of the different approaches to diatom-based paleoenvironmental reconstructions.

# ABSTRACT

Diatoms are an important and often dominant component of the microalgal assemblages in estuarine and shallow coastal environments. Given their ubiquity and strong relationship with the physical and chemical characteristics of their environment, they have been used to reconstruct paleoenvironmental changes in coastal settings worldwide. The quality of the inferences relies upon a deep knowledge on the relationship of modern diatom species and their ecological requirements, as well as on the taphonomic constrains that can be affecting their preservation in sediments. In Argentina, information on estuarine diatom ecology is scattered and fragmentary. Studies on estuarine diatoms from the $20^{th}$ century have been mostly restricted to taxonomic descriptions of discrete assemblages. Given the lack of detailed studies on the distribution of modern diatoms in local estuarine environments and their relationship with the prevailing environmental conditions, most paleoenvironmental reconstructions were based on the ecological requirements of European diatoms. However, studies on diatom distribution along estuarine gradients from Argentina have increased in recent years, constituting a potential source of data for paleoecologists. In this chapter, the literature on modern estuarine diatoms from Argentina is revised in order to synthesize the available ecological information and to detect possible modern analogues for Quaternary diatom assemblages. The main objective is to build bridges between ecology and paleoecology, and to discuss the reaches and limitations of the different approaches to diatom-based paleoenvironmental reconstructions. Further studies exploring the relationship between estuarine diatom distribution and environmental characteristics are necessary in order to increase the precision of paleoenvironmental inferences in the region and to generate new hypothesis for further study.

# INTRODUCTION

Estuaries are transitional environments located between rivers and the sea, characterized by widely variable and often unpredictable hydrological, morphological and chemical conditions (Day et al., 1989). Given these particular environmental characteristics, estuarine organisms are often restricted to limited sections of estuarine gradients, resulting in well-developed distribution patterns (Moore & McIntire, 1977; Ysebaert et al., 2003; De Francesco & Isla, 2003).

Diatoms are the main source of primary production in shallow estuarine systems (Admiraal, 1984; Colijn et al., 1987; Wolfstein et al., 2000; Rybarcyk & Elkaïn, 2003), serving as an essential supply of food for numerous species of zooplankters and deposit feeders (Bianchi & Rice, 1988; Bennett et al., 2000; Rzeznik-Orignac et al., 2003) and forming biofilms that increase the resistance of sediment surface to erosion (Paterson, 1989; Underwood & Paterson, 1993; Underwood, 1997; Austen et al., 1999; Bergamasco et al., 2003). Laboratory experiments showed that different diatom species have different levels of tolerance to salinity, nutrient concentrations, temperature and light availability (Admiraal, 1977a,b,c,d; Admiraal & Peletier, 1980; Admiraal et al., 1982). Distribution patterns observed in the field usually respond to a combination of these variables (Moore & McIntire, 1977; Amspoker & McIntire, 1978; Oppenheim, 1991; Underwood, 1994; Gómez et al., 2004). Moreover, the distribution of diatoms in estuarine environments is the result of a complex set of interactions between environmental variables and interspecific competitive interactions (Underwood, 1994).

Given their sensitivity to environmental variables and abundance in sediments, diatoms constitute useful indicators for the study of paleoenvironmental changes (Cooper, 1999). This has been well known since the late 1890s, when the pioneering studies of Cleve (1894/1895) demonstrated that benthic diatom assemblages from surface sediments reflect the physical and chemical characteristics of the overlying water masses (Maynard, 1976). However, only after the 1920s the value of diatom analysis in paleoenvironmental reconstructions was recognized (Denys & De Wolf, 1999). Since the identification of salinity as a major determinant of diatom distribution, the remains of these organisms have become widely used as paleoenvironmental indicators in coastal deposits. Furthermore, a variety of problems in coastal geology were tackled by applying diatom-based methods, covering fields such as stratigraphy, coastal processes, paleogeography, sea-level and climate changes (Denys & De Wolf, 1999). In estuarine systems, they have also been used to define the naturally occurring state of the ecosystem, in order to infer historical changes due to human influences (Cooper, 1999).

The methods used in paleoenvironmental reconstructions rely on the general assumption that the environmental requirements of the fossils used as bioindicators have remained constant during the period considered and, consequently, are similar to those of their closest living representatives. In this way, the environmental information obtained from living organisms can be used as modern analogous and extrapolated to the fossil record, particularly in Quaternary research. This approach is based on a strict substantive application of the principle of Taxonomic Uniformitarianism (*the ecology of modern organisms is the key to that of past organisms*; Dodd & Stanton, 1990). Estuarine diatom-based paleoenvironmental reconstructions have been based in autoecological or synecological techniques. In autoecological studies, the composition of modern diatom assemblages is analyzed, and relevant environmental requirements of each species or group of species are considered (De Wolf, 1982; Vos & De Wolf, 1988, 1993; Denys & De Wolf, 1993). In the last decades, the great volume of autoecological data available for European diatoms has been summarized as a series of ecological codes (De Wolf, 1982; Vos & De Wolf, 1988, 1993; Denys, 1991/1992; Van Dam et al., 1994). The most commonly used diatom classifications in coastal areas were based on salinity tolerances (polihalobous, mesohalobous, oligohalobous halophilous, oligohalobous indifferent and halophobous; Hustedt, 1953) and life forms (plankton, epiphytes, benthos, and aerophilous; De Wolf, 1982). Later, Vos & De Wolf (1988) combined both classifications in order to define autoecological groups (i.e. marine/brackish epiphytes, brackish/freshwater tychoplankton) characteristic of different coastal habitats. Specific sedimentary environments in coastal wetlands were characterized on the basis of the relative frequencies of the 16 ecological groups defined (Vos & De Wolf, 1988).

Besides its usefulness, the application of autoecological techniques to the interpretation of past environmental changes has limitations and needs to be interpreted with caution. This methodology is based on the classification of single taxa in autoecological categories delimited by general ecological borderlines (Vos & De Wolf, 1993). Although to some extent such borders can be drawn, there are many cases of gradual species turnover along environmental gradients in nature, and many taxa have large adaptability to changing environmental conditions (Denys & De Wolf, 1999). This is particularly true for estuarine environments, where most taxa usually show wide salinity tolerances, making it difficult their placement into discrete categories (Licursi et al, 2006; Hassan et al., 2009). In fact, this difficulty of assigning a taxon unambiguously to an individual class constitutes one of the main problems of the autoecological classification (Battarbee et al., 1999).

In contrast to the use of generalized autoecological concepts, synecological techniques are based on the application of statistical inference models derived from modern contemporaneous species-environment relations, allowing quantitative inference of important parameters. A set of regional observations seems imperative in this, since hydrographic and ecological conditions differ between study areas (Denys & De Wolf, 1999). The statistical calibration of selected environmental variables and dead diatom assemblage composition (*transfer functions*) constitutes the most precise method, since it is based on the study of the entire diatom assemblage rather than on individual taxa (Juggins, 1992; Ng & Sin, 2003). In the last decades the need for quantification in Quaternary research has increased and a great number of diatom-based transfer functions have been developed in coastal and estuarine environments of the Northern Hemisphere (Juggins, 1992; Campeau et al., 1999; Sherrod, 1999; Zong & Horton, 1999; Gehrels et al., 2001; Ng & Sin, 2003; Sawai et al., 2004; Horton et al., 2006).

In Argentina, information on estuarine diatom ecology is scattered and fragmentary, and there is a lack of detailed distributional studies. As most diatom taxa are cosmopolitan, the autoecological information necessary to carry out local paleoenvironmental reconstructions has been historically gathered from European datasets (e.g. Espinosa, 1998, 2001; Espinosa et al., 2003). Studies on modern estuarine diatoms from Argentina during the $20^{th}$ century have been mostly restricted to taxonomic descriptions of discrete assemblages (see Vouilloud, 2003). Works on diatom distribution along estuarine gradients have increased during the $21^{th}$ century, constituting a potential source of data for paleoecologists. However, the information provided by these ecological studies has not always been applied to infer paleoenvironmental conditions from fossil diatoms. This points to the question if the lack of contact between paleoecological and ecological studies may be responding to methodological barriers between both disciplines rather than to a real scarcity of information.

In this chapter, the literature on modern estuarine diatoms from Argentina is reviewed in order to summarize the available ecological information and to evaluate its usefulness as modern analogues for Quaternary diatom assemblages. The main objective is to build bridges between ecology and paleoecology, and to discuss the reaches and limitations of the different approaches to diatom-based paleoenvironmental reconstructions. Although the discussion will focus on estuarine settings from Argentina, it could be useful for guiding the debate in other regions or environmental settings with similar research histories. The main questions to be addressed are: 1) Do estuarine diatoms reliably reflect estuarine environmental conditions? 2) How much information about ecological requirements of estuarine diatoms do we have? 3) How can researchers improve the quality of diatom-based paleoenvironmental inferences in coastal settings?

*Chapter 1*

# DO ESTUARINE DIATOMS RELIABLY REFLECT ESTUARINE ENVIRONMENTAL CONDITIONS?

The first issue to take into account in order to transfer ecological information to the past is to understand how accurately fossil organisms reflect their living environment and how much environmental information become lost in their transition from live to dead assemblages. This subject is particularly essential in the study of sedimentary diatom assemblages, since they are the result not only of ecological processes that drive the distribution of living diatoms along the environmental gradients, but also of taphonomic processes (i.e., the postmortem history of dead remains) that alter dead frustules after their deposition. Therefore, their distribution within a locality may not necessary constitute an accurate representation of their living habitat (Juggins, 1992; Vos & De Wolf, 1993; Sherrod, 1999). In highly variable and energetic environments, such as coastal and estuarine areas, taphonomic processes can so drastically alter the species composition of a diatom assemblage that the original ecological signals reflected by the *in situ* assemblage may be either obscured or obliterated (Sherrod, 1999). Thus, the assessment of how accurately dead diatom assemblages preserve the original environmental information becomes a main requisite in order to evaluate the applicability of modern data sets.

When looking for modern analogues of paleoenvironments, most researchers turn to the surface sediment diatom thanatocoenoses (dead diatoms, both autochthonous and allochthonous remains, present at a particular place in the sediment; Sherrod, 1999), which are assumed to integrate small-scale temporal and spatial perturbations into more defined assemblages; consequently, they are assumed to be more accurate indicators of general environmental conditions than biocoenoses (living communities). The use of diatom thanatocoenoses as modern analogous is based on the general assumption that dead diatom assemblages faithfully reflect the environmental conditions prevailing at the sampling point. Hence, they are considered reliable indicators of environmental parameters, without requiring time consuming seasonal studies (Juggins, 1992).

The most common approach to the evaluation of the ecological fidelity of fossil assemblages has been the testing of agreement between living communities and the locally accumulating dead assemblages in modern environments. This method has led to powerful guidelines for paleoecological reconstruction in foraminifers (e.g., Goldstein & Watkins, 1999; Horton, 1999; Murray & Pudsey, 2004), ostracodes (Alin & Cohen, 2004), mollusks and brachiopods (e.g., Kidwell, 2001, 2002; Kowalewski et al., 2003). However, there is a general lack of detailed quantitative works attempting to evaluate the fidelity of coastal diatom assemblages.

In Argentina, this approach has been recently applied by Hassan et al. (2008), who analyzed the environmental fidelity of dead diatom assemblages along two microtidal estuaries (Mar Chiquita coastal lagoon and Río Quequén Grande; Fig. 1) and discussed their potential use as modern analogues in paleoenvironmental reconstructions. A good agreement between live benthic communities and total surface assemblages was found in both estuaries. The comparison between live cells and empty frustules did not allow the recognition of a significant allochthonous component. Although relatively high percentages of empty frustules were found in the tidal inlet zone from Mar Chiquita coastal lagoon, they originated mainly from taxa found alive in the same site. Similar results were obtained in tidal flats from salt marshes of Japan, where only 3% of the empty frustules present in surface sediments of the littoral zone were found to be allochthonous (Sawai, 2001). The investigation about possible and net effects of transport on population composition in other groups, led to the general conclusion that out-of habitat postmortem transport does not constitute an overwhelming taphonomic problem in ordinary depositional settings (Kidwell & Flessa, 1995; Horton, 1999; Behrensmeyer et al., 2000; Alin & Cohen, 2004). These results, together with the good preservation shown by diatom valves, suggest that benthic diatom assemblages are not under significant alteration by biostratinomic and early-diagenetic processes along the estuarine foreshore: although mixing of autochthonous and allochthonous diatoms does occur, estuarine dead diatom assemblages still reflect the environmental gradient with high fidelity. As a consequence, they constitute useful modern analogues for paleoenvironmental reconstructions and provide advantages over the use of live communities. Moreover, since paleoecologists have only total sedimentary assemblages available to examine and interpret (Scott & Medioli, 1980), the understanding of taphonomic alterations suffered by them leads to an increase in the precision of paleoenvironmental interpretations.

In contrast to the use of benthic diatoms, the application of modern ecological data gathered from phytoplanktonic assemblages becomes a more problematic issue. According to a strict definition (Birks & Birks, 1980) the term allochthonous refers to those individuals transported away from their life position before burial. It has been proposed that only benthic taxa should be considered autochthonous and used in palaeoecological reconstruction, since plankton forms are by definition allochthonous and, thus, more subject to lateral transport by tides and currents (Simonsen, 1969). Vos & De Wolf (1993) also emphasized life form as an important variable to interpret paleoenvironments, pointing out that marine plankton and tychoplankton diatoms are basically allochthonous components, whilst epiphytic and epipsammic diatoms are probably autochthonous. Accordingly, a wide distribution of empty valves and frustules of the tychoplanktonic *Paralia* sp. was observed throughout the entire tidal zone in marshes from Japan as a consequence of their transport by currents action (Sawai, 2001).

The representation of phytoplanktonic diatom species in surface sediments of Argentinean estuaries has not been systematically assessed. Frenguelli (1935, 1941) remarked the large differences in the salinity tolerances of diatom assemblages of sedimentary and plankton net samples from Río de la Plata and Mar Chiquita estuaries (Fig. 1), which were attributed to taphonomic biases (see Río de la Plata and Mar Chiquita sections below). Licursi et al. (2006) recorded up to 70% of empty frustules in plankton net samples from Río de la Plata, which were closely related to bathymetry. These high percentages of empty frustules belonged mainly to freshwater diatoms, which were probably allochthonous riverine elements transported from the headwaters (Gómez et al., 2004; Licursi et al., 2006). High percentages of tychoplanktonic taxa were found in sediment samples from Mar Chiquita and Río Quequén Grande estuaries, but as their distribution along the estuarine gradient was consistent with their salinity tolerances, they were not ecologically out of place (Hassan et al., 2008). Moreover, as tychoplanktonic diatoms are closely associated to the sediment, they are less prone to lateral transport than true plankton. To sum up, systematic studies comparing the diatom assemblage composition in surface sediments and the overlying water column are needed in order to estimate their grade of preservation and environmental fidelity. Meanwhile, caution is needed when paleoenvironmental inferences in estuaries are derived from phytoplanktonic diatom assemblages.

Figure 1. Location map showing the main Argentinean estuaries.

*Chapter 2*

# HOW MUCH INFORMATION ABOUT ECOLOGICAL REQUIREMENTS OF ESTUARINE DIATOMS DO WE HAVE?

The Argentina coastline has a wide variety of estuaries ranging from the widest in the world (Río de la Plata) to very small ones located in areas of very difficult access (Fig. 1). Due to the different climates that characterize the Argentinean territory, the estuaries show different discharges, being the Río de la Plata the largest one. Tidal amplitudes also vary significantly, being microtidal between the Río de la Plata and the Río Quequén Salado, mesotidal in the coast between Bahía Blanca estuary to Río Chubut, and macrotidal along the rest of the Patagonian estuaries (Piccolo & Perillo, 1999).

Vouilloud (2003) published a review listing of the publications about Argentinean diatoms from the 19th century to the '90 decade. Of the revised literature, only a small proportion of the articles (see figure 2 in Vouilloud, 2003) dealt with modern estuarine diatoms. Among them, taxonomic studies were the most numerous, although some ecological articles (mainly focused on the whole phytoplanktonic assemblage) were also published. Only recently, some distributional studies on estuarine diatoms were published, particularly for Río de la Plata (Licursi et al., 2006; Gómez et al., 2009); Mar Chiquita coastal lagoon (Hassan et al., 2006; 2009), Río Quequén Grande and Río Quequén Salado (Hassan et al., 2007; 2009).

In the following sections, the state of the knowledge on each of the main estuaries from Argentina is reviewed, focusing mainly on the ecological requirements of the dominant diatom taxa, and stressing the value of the information presented for paleoecological purposes. In order to summarize the available information, tables listing all the reviewed works of recent publication (Appendix I) and the diatom taxa cited in them (Appendix II) were constructed. Comprehensive lists of the diatom taxa cited in older works can be found in Ferrando et al. (1962), Ferrario and Galván (1989), Vouilloud (2003) and Sar et al. (2009). In order to allow the comparison of data among the different reviewed works, all diatom names and their authorities were updated to their currently accepted name following Algaebase (Guiry & Guiry, 2009) and WoRMS (SMEBD, 2009) taxonomic databases.

## RIO DE LA PLATA ESTUARY

The Argentina coast starts in the Río de la Plata estuary (Fig. 1), located at about 35°S on the Atlantic coast of South America. The river drains the second largest basin of this continent, following that of the Amazon (Piccolo & Perillo, 1999). Its drainage area covers ca. $3.1 \times 10^6$ km$^2$, which represents about 20% of the South American continental area (Acha et al., 2008). It forms one of the most important estuarine environments in South America, being a highly productive area that sustains fisheries in Uruguay and Argentina. The estuary is characterized by a salt-wedge regime, low seasonality in the river discharge, low tidal amplitude (<1m), a broad and permanent connection to the sea, and high susceptibility to atmospheric forcing, due to its large extension and shallow water depth (Acha et al., 2008 and references therein).

The Río de la Plata estuary and its oceanic front has been the most extensively studied of Argentina. The first phytoplanktonic diatom from the Río de la Plata, *Caloneis bivittata* var. *rostrata*, was mentioned by Heiden (Schmidt et al., 1874-1959). Tempère and Peragallo (1907-1915) mentioned 8 new forms. The list increased to 68 forms during the 1920s and the 1930s, with a series of taxonomic studies which focused on plankton samples of the estuary (Carbonell & Pascual, 1924; Hentschel, 1932; Thiemann, 1934; Carbonell, 1935; Cordini, 1939).

Frenguelli (1941), studied 3 plankton and 1 bottom sediment samples collected in three different points of the estuarine gradient (inner estuary, middle estuary and mouth, Fig. 2A). A total of 309 taxa, present at very low abundances, were observed. The dominance of these taxa, mostly benthic, epiphytic and aerophilic forms, was related to the transport of littoral diatoms from the headwaters and the adjacent coast. The assemblage composition of the plankton samples was homogeneous and characteristic of estuarine environments. They were dominated by *Aulacoseira granulata* and *A. ambigua*, accompanied by some freshwater and marine taxa (Fig. 3). The bottom sediment sample, on the other hand, showed very scarce diatom frustules, mostly marine neritic forms, with only one species (*Paralia sulcata*) classified as frequent. Detailed taxonomical descriptions of the dominant taxa were provided, together with information on their ecological preferences. The later data, however, were taken from European floras (particularly Hustedt 1937/1938), and no *in situ* measurements of the main environmental parameters from the sampling site were provided. Guarrera (1950) analyzed the composition of the phytoplanktonic assemblage in two sampling stations located near Buenos Aires city, identifying 16 genera. Müeller Melchers (1945, 1952, 1953, 1959) worked on plankton samples from the Río de la Plata maritime front, listing and providing taxonomic descriptions for 69 diatom taxa. Although the number of studies on phytoplankton increased significantly since the 1970s, most studies focused on the coastal areas and maritime front (Balech, 1976, 1978; Martínez Macchiavelo, 1979; Lange, 1985; Baysee et al., 1986; Elgue et al., 1990; EcoPlata Team, 1996; Gayoso, 1996), being scarce studies on the estuarine zone of the river (Roggiero, 1988, CARP-SIHN-SOHMA, 1989).

Many works on the composition and dynamics of the phytoplankton have been carried out in the estuarine zone of the river in the last decades ( Gómez & Bauer, 1998a, 1998b, 2000; Cervetto et al., 2002; Gómez et al., 2002, 2004; Carreto et al., 2003, 2008; Calliari et al., 2005, 2009). In most of these studies diatoms represent one of the dominant groups. The centric taxa *Aulacoseira granulata* var. *angustissima, A. granulata, A. distans, A. ambigua, Actinocyclus normanii, Thalassionema nitzschioides, Stephanodiscus hantzschii* and *Skeletonema costatum* were mentioned among the dominant diatoms in most of these studies. The dominance of these taxa was explained as a consequence of their capability for exploiting this low light environment owing to their efficient light-harvesting mechanisms (Gómez & Bauer, 1998). Unfortunately, although lists of the dominant taxa and environmental information are provided, none of these works presents information on the patterns of distribution of each taxon along the estuarine gradient or their ecological preferences. Hence, the way in which the data are presented limits their usefulness for paleoenvironmental applications.

From an autoecological point of view, the most valuable information on phytoplanktonic diatom ecology and distribution in the Río de la Plata estuary was provided by Licursi et al. (2006), who studied the factors affecting the composition and structure of diatom phytoplankton across the estuarine gradient. Samples were collected with plankton nets from 29 sites distributed along a gradient of estuarine conditions from the headwaters to the estuary mouth. The estuarine gradient was divided into 6 zones of 50 km long and sites grouped according to them (Fig. 2A). For each zone, data on environmental variables were also provided (Fig. 2B). As reported in previous works, the assemblages were dominated by chains of centric diatoms (Fig. 3). Canonical Correspondence Analysis (CCA) was performed in order to relate diatom assemblages to environmental variables, allowing recognizing two groups of taxa: the first group was related to low values of salinity, pH and concentrations of dissolved oxygen and higher amounts of suspended solids and nutrients (sections A to C, Fig. 3). The second group clustered taxa that tolerate higher salinity and alkalinity (sections D to F, Fig. 3). This assemblage was characteristic of marine environments, and had a lower limit of salinity tolerance of 7-8. No taxa exclusive of brackish waters were identified, but some freshwater and marine taxa presented wide salinity tolerances. Despite the taphonomic limitations to the use of phytoplanktonic taxa as modern analogues in estuarine environments, the information on diatom distribution and environmental preferences provided in this work is of great utility for coastal paleoenvironmental reconstructions.

Microphytobenthic diatom communities from Río de la Plata, on the other hand, have received little attention. Metzeltin and García-Rodríguez (2003) published a book on the taxonomy of the Uruguayan diatoms based on the analysis of samples of periphyton collected along the Uruguayan coast of the estuary, listing and illustrating 295 species. Bauer et al. (2007) assessed the usefulness of biofilms covering *Schoenoplectus californicus* (a bulrush widely distributed along the shore of the Río de la Plata) as indicators of water quality. They selected three sampling sites in the freshwater tidal zone of the estuary (salinity <0.5) subjected to different grades of human impact and analyzed the taxonomic composition and tolerances of the taxa present over *S. californicus* stems. Diatoms constituted one of the dominant organisms in the biofilms, and their distribution was mainly conditioned by turbidity, pH, salinity and water-quality variables. Two assemblages were defined: one related to the highest turbidity values (average 50±22 NTU), dissolved oxygen (average 7.5±1 mg l-1) and pH (average 7.4±0.5), and included pollution sensitive species such as *Encyonema silesiacum, Navicula erifuga, N. rynchocephala, Neidium dubium, Nitzschia fonticola, N. nana, Placoneis clementis* and *Pleurosira laevis*, and less tolerant species such as *Gomphonema augur, Luticola ventricosa* and *Nitzschia brevissima*. The second assemblage was related to high conductivities (average 740±200 µS cm-1), ammonia (average 1.7±1.1 mg l-1), nitrates (average 0.08±0.04 mg l-1) and phosphates (average 0.87±0.42 mg l-1) concentrations. This group included mainly taxa characteristic of polluted sites such as *Nitzschia palea*.

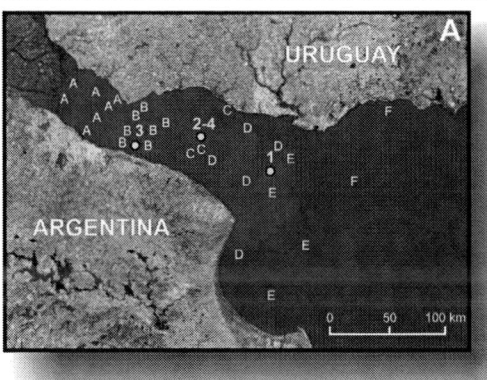
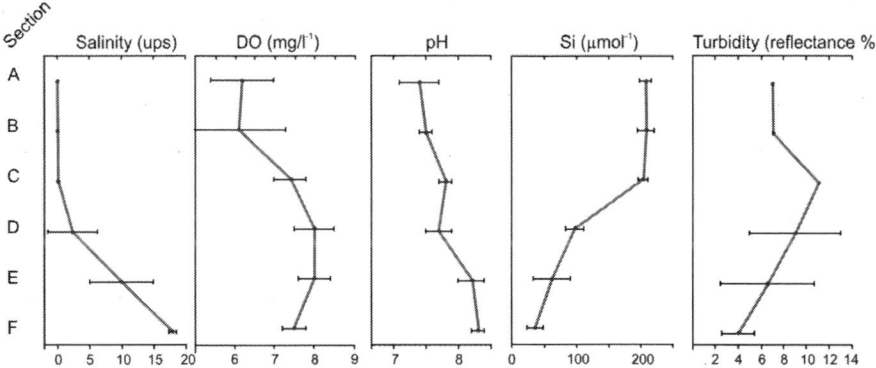

Figure 2. A) Location of sampling sites from Frenguelli (1941, numbers) and Licursi et al. (2006, letters) at the Río the la Plata estuary, and B) Summary of environmental information provided by Licusi et al (2006).

In a recent contribution, Gómez et al. (2009) analyzed the seasonal and spatial distribution of microbenthic communities in 10 sites located along 155 km of the estuarine shoreline. Diatoms were abundant, particularly during autumn. *Navicula novaesiberica, N. erifuga, Fallacia pygmaea, Nitzschia palea, Amphora lybica* and *Sellaphora pupula*, were the most abundant taxa (>60%). According to their relationship with environmental variables, the whole assemblage was separated into two groups by CCA: the first group was composed by *L. ventricosa, Stauroneis brasiliensis* and *Fallacia omissa*, and related with the highest nitrite (0.14±0.10 mg $l^{-1}$) and ammonia (0.30±0.23 mg $l^{-1}$) values. The second group of species included *Amphora acutiuscula, A. lybica, Pleurosira laevis, Actinocyclus normanii, Staurosirella pinnata, Hantzschia amphioxys, Hippodonta hungarica*, and *Navicula tenelloides*, associated with high conductivity (1657 ± 1597 µS $cm^{-1}$), and *Nitzschia lacunarum* linked to high concentrations of nitrates (0.94 ± 0.17 mg $l^{-1}$). Although this study covered a large portion of the estuarine gradient and provided detailed environmental data for each sampling point, the information on the distribution of single taxa in each sampling station was not presented. Hence, it is not possible to extract information on single taxa environmental preferences, which would be very useful in autoecological paleoenvironmental reconstructions.

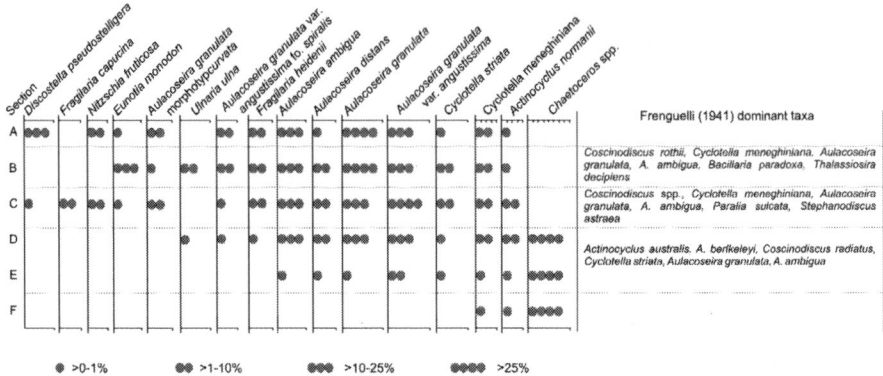

Figure 3. Relative frequencies of diatom assemblage composition for Río de la Plata estuary (based on data from Licursi et al., 2006). Dominant taxa found by Frenguelli (1941) in each section are listed.

## Mar Chiquita Coastal Lagoon

Mar Chiquita is the only coastal lagoon of Argentina that is chocked with a long inlet (Piccolo & Perillo, 1999). It is a brackish-water body with a surface area of 46 km$^2$ and a mean depth of 0.6 m, extending along the microtidal Argentinean coast (Figs. 1 and 4). From a hydrological point of view, the coastal lagoon can be divided into an innermost shallow zone, where the tidal effect is not significant, and an estuarine channel subjected to tidal action (Reta et al., 2001). Sediments are mainly composed of sand and silt with high proportions of mollusk shells. The shallow depth and particular dynamics of the coastal lagoon induces sediment reworking and prevents the development of a stable salinity gradient (Fasano et al., 1982). Nutrients and suspended sediment concentration are higher in the inner areas of the lagoon than in the tidal channel, whereas salinity, current speed and depth show the opposite pattern (Schwindt et al., 2004).

The study of diatoms from Mar Chiquita began with Frenguelli (1935) who described the assemblages present in two samples collected from the inlet of the coastal lagoon (Table 1). The first was a sediment sample taken from the bottom of the inlet, which contained relatively scarce diatom frustules of marine-neritic origin. In the second sample, which was taken with plankton net, diatom frustules were conspicuous, and consisted in a mixed assemblage of fluvial, lacustrine and estuarine taxa, characteristic of both oligohaline and mesohaline conditions. The significant differences between both assemblages were taphonomicaly explained: whereas than in the plankton sample diatom assemblages reflected an average of the living communities that succeeded in the very changing ecological environment, the diatom composition of the sediment sample was interpreted as a reworked fossil assemblage which indicates that in the past the zone was a marine bay (Frenguelli, 1935).

**Table 1. Diatom assemblage composition and environmental significance of the two samples collected by Frenguelli (1935) at Mar Chiquita coastal lagoon**

| | Bottom sample | Plankton net sample |
|---|---|---|
| Abundant species | *Paralia sulcata* | ------------ |
| Frequent species | *Actinocyclus vulgaris* | *Aulacoseira granulata* <br> *Bacillaria paradoxa* <br> *Cyclotella meneghiniana* <br> *Navicula peregrina* <br> *Nitzschia circumsuta* <br> *Tropidoneis lepidoptera* var. *proboscidea* |
| Ecological Conditions | Marine/neritic assemblage. Fossil and reworked, probably indicating the presence of a marine bay in the past. | Mixed assemblage of mesohalobous and oligohalobous taxa, of lacustrine, fluvial and estuarine origin. Planktonic and benthic. Assemblage composition reflects the mean environmental conditions of the basin. |

No new studies on Mar Chiquita diatoms were conducted until the $21^{st}$ century. Recently, the temporal and spatial dynamics of the phytoplankton and its relation to nutrient concentrations were studied (De Marco, 2002; De Marco et al., 2005). Although diatoms constituted the dominant assemblage, taxa were identified only at the genus level.

Espinosa et al. (2006) analyzed the distribution of surface diatom assemblages across the marsh in a sampling station located in the Mar Chiquita tidal inlet (site 6, Fig. 4A). The marsh was divided into five subenvironments: floodplain, distant and closer high marshes, levee/chenier, and mudflat. In the flood plain, where tidal submersion is infrequent and of short duration, the assemblage was dominated by brackish/epiphytic and aerophilous taxa (Fig. 5). Brackish/freshwater epiphytic and tychoplanktonic diatoms dominated the distant high marsh, whereas the closer high marsh was dominated by the brackish aerophilous *Diploneis interrupta,* a taxon typical of supratidal environments. The levee and chenier zone, where the tidal flooding is frequent, was dominated by marine planktonic, benthic and epiphytic taxa. The diatom assemblage of the mudflat was dominated by a mixture of marine (epiphytic and benthic) and freshwater (planktonic and tychoplanktonic) taxa. Overall, the composition of diatom assemblages in this microtidal marsh was related to morphology, duration and frequency of tidal exposure, and the consequent salinity fluctuations.

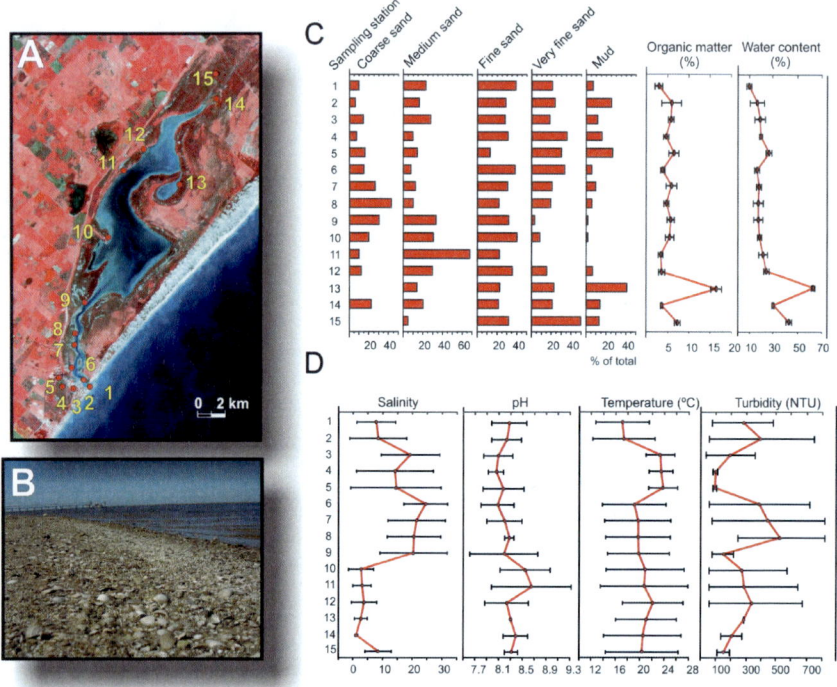

Figure 4. Location of sampling sites (A), view of the estuarine zone (B), and the corresponding sedimentary (C) and water quality (D) parameters at Mar Chiquita coastal lagoon (modified after Hassan, 2008).

Hassan et al. (2006) studied the diatom assemblages dominating in surface sediments along a transect from the inlet to the inner reaches of the coastal lagoon in relation to the main environmental parameters (Fig. 4). Most diatom species found were highly euryhaline taxa, adapted to the great salinity and tidal ranges that characterize the lagoon. Besides salinity, other environmental factors such as turbidity, temperature and sediment properties were important in explaining diatom assemblage composition. The marine/brackish diatoms *Catenula adhaerens* and *Opephora pacifica* dominated in the tidal channel, whereas the inner lagoon was dominated by the brackish/freshwater tychoplanktonic diatoms

*Staurosira venter* and *Staurosirella pinnata* (Fig. 6). Similar distributional patterns, characteristic of environments with fluctuating salinity regimes, have been observed in other coastal lagoons from the Atlantic Ocean coasts (e.g., Sylvestre et al. 2001; Bao et al., 2007; Witkowski et al., 2009). In these environments, taxa are selected according to their ability to adapt to changing salinity rather than to their salinity optima (Snoeijs, 1999). The diatom assemblages from Mar Chiquita coastal lagoon are of particular importance for the paleoenvironmental reconstruction of the many estuarine lagoons developed in the microtidal Argentinean coast during the Holocene marine transgression (Espinosa et al., 2003; Hassan et al., 2009).

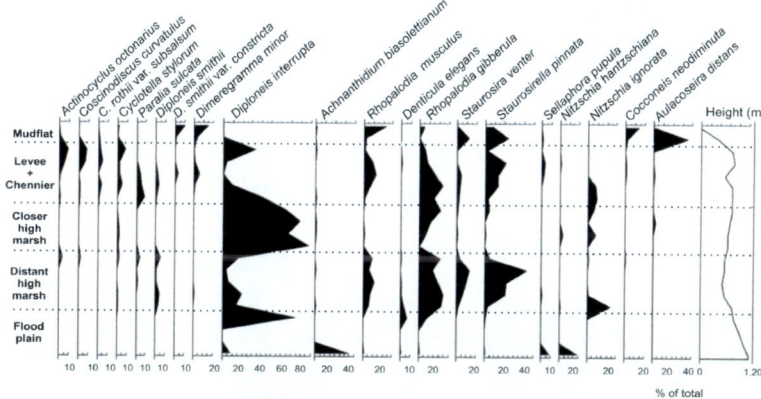

Figure 5. Distribution of the dominant diatom taxa across the Mar Chiquita lagoon marsh (modified after Espinosa et al., 2006).

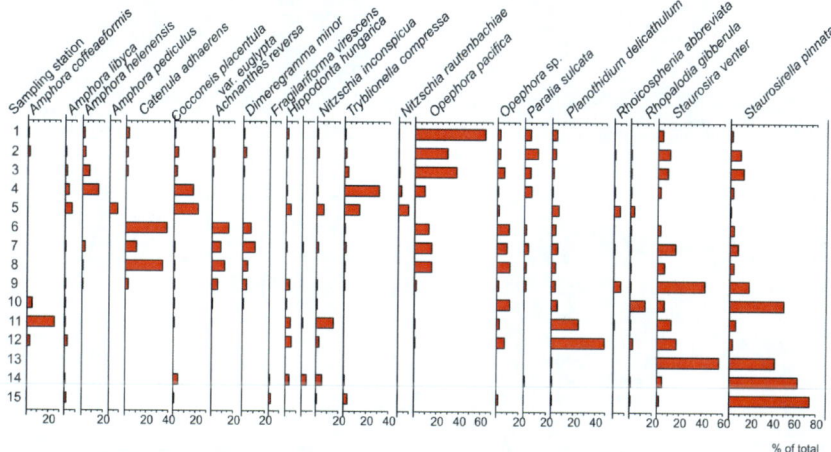

Figure 6. Diatom assemblage composition at Mar Chiquita coastal lagoon (modified after Hassan et al., 2009).

## QUEQUEN GRANDE ESTUARY

The Río Quequén Grande is a partially mixed estuary discharging at the microtidal coastline of northern Argentina (Figs.1 and 7). Mean depth is 2–3 m and width is 150–200 m. Most of the river runs on Pleistocene partly cemented loessic sediments. Due to the sediment characteristics – silty loess with caliche levels – large portions of the river flow within a canyon whose walls reach up to 12 m high (Perillo et al., 2005). There is no significant accumulation of sediment on the bottom, and the river is well known by its rapids, composed of indurate levels of caliche. However, the river carries large amounts of silt during floods. Salinity decreases significantly along the estuarine gradient, the highest salinities (20–25) are found within the first 2–3 km of the inlet; approximately 10 km upstream, salinity decreases to 0–1 (Fig. 7C). Given its economic and strategic importance, the estuary has been the focus of many man-made modifications (i.e., dredging, jetty and harbour construction, etc.) that have reduced water circulation producing strong reductive and even anoxic conditions (Perillo et al., 2005).

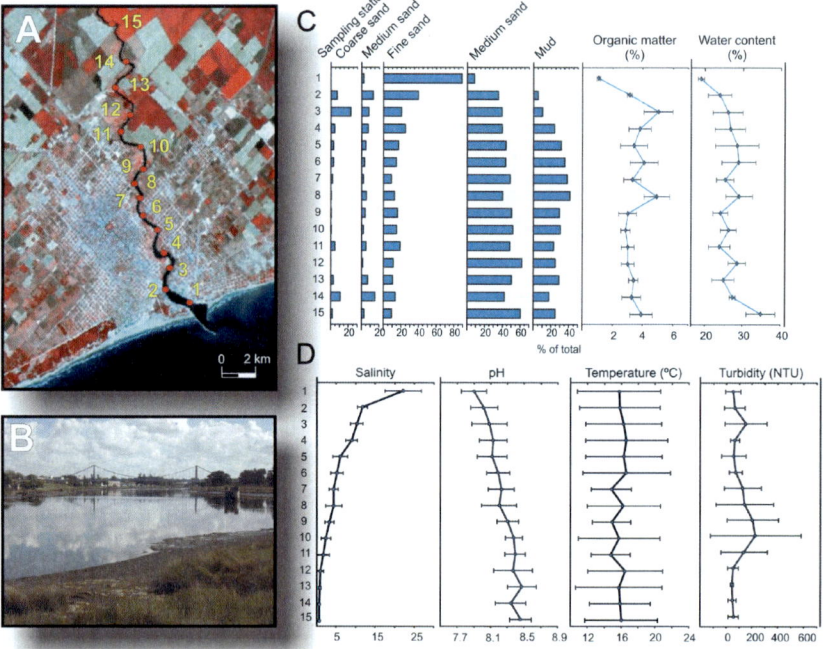

Figure 7. Location of sampling sites (A), view of the estuarine zone (B), and the corresponding sedimentary (C) and water quality (D) parameters at Quequén Grande river (modified after Hassan, 2008).

The composition of the diatom assemblages present in surface sediments from the estuary have been recently studied (Hassan et al., 2006, 2007). Diatom composition was significantly related to salinity, and the assemblages showed gradual turnovers along the stable salinity gradient that characterizes the estuary. The marine/brackish diatoms *Amphora helenensis* and *Opephora pacifica* dominated in the inlet, while the brackish/freshwater diatoms *Cocconeis placentula* var. *euglypta* and *Nitzschia inconspicua* increased their relative frequencies towards the middle estuary. A diverse freshwater assemblage, characterized by *Achnanthidium minutissimum*, *Amphora pediculus*, *Hippodonta hungarica*, *Denticula kuetzingii* and *Rhoicosphenia abbreviata*, dominated the upper estuary (Fig. 8). Similar diatom zonations were recorded in estuaries characterized by stable salinity gradients (Moore & McIntire, 1977; Ampsoker & McIntire, 1978; Juggins, 1992; Debenay et al., 2003; Resende et al., 2005). As salinity is one of the main environmental factors controlling diatom distribution in estuaries (Cooper, 1999), the diatom zonation observed in the Quequén Grande estuary was explained by the existence of a stable salinity gradient. Hence, the strong relationship between diatoms and salinity in the estuary makes them useful analogues for inferring past salinity changes in the region.

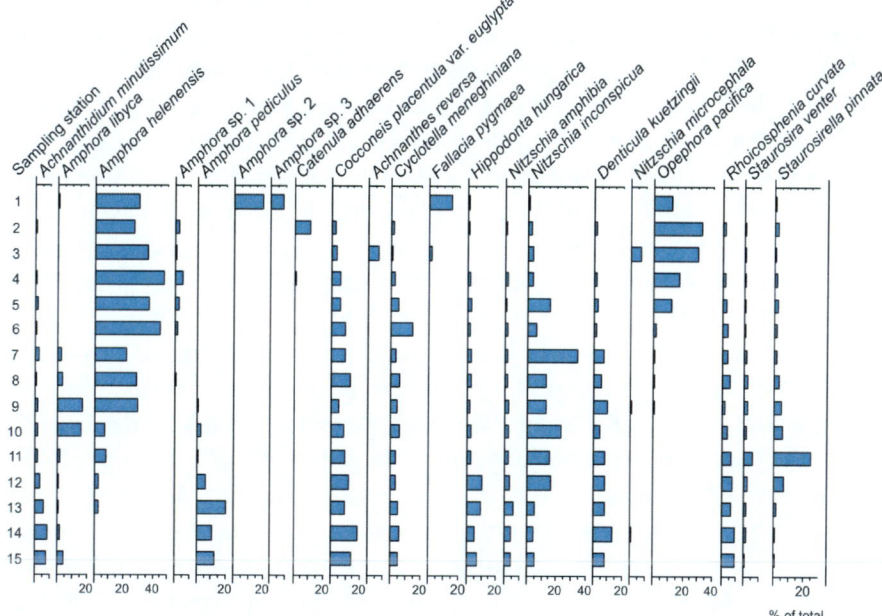

Figure 8. Diatom assemblage composition at Quequén Grande river (modified after Hassan et al., 2009).

## QUEQUEN SALADO ESTUARY

The Río Quequén Salado is located 100 km westwards of the Río Quequén Grande, constituting the northernmost estuary subjected to a mesotidal regime in the Argentinean coast (Figs. 1 and 9). The estuary has been minimally impacted by human activity because of the absence of large urban settlements, bridges, jetties or harbors. Moreover, it has been suggested that, although a bit smaller, the Quequén Salado estuary presently represents similar conditions to those of the Quequén Grande estuary prior to the anthropogenic influence (Perillo et al., 2005). The lower valley is oriented to the SSE, with steep walls of 8–15 m high. This portion of the river is also characterized by rapids caused by resistant caliche levels. In the last 5 km the river runs across a sandy barrier composed of vegetated dunes (Marini & Piccolo, 2000).

The study of surface sediment diatom assemblages from Río Quequén Salado estuary, which started very recently, yielded distributional patterns very similar to those found in Quequén Grande estuary, as both present stable salinity gradients (Hassan et al., 2007). Marine and marine/brackish diatoms, such as *Paralia sulcata*, *Cymatosira belgica* and *Amphora helenensis*, dominated the lower and middle estuary, and were gradually replaced by the brackish/freshwater and freshwater taxa *Nitzschia inconspicua* and *Hippodonta hungarica* towards the headwaters (Fig. 9). However, the marine/brackish diatom assemblage was more widely distributed in Río Quequén Salado and had no analogues when compared to the assemblages represented in Quequén Grande. This difference between both estuaries may be related to differences in salinity and grain size distribution. In fact, the range of salinities and sediment grain sizes in the first kilometers of the Quequén Salado estuary were higher than those recorded at Quequén Grande estuary, where polyhaline conditions and sandy sediments were recorded only in the first meters of the inlet. The differences between both estuaries were attributed to the tidal range and the grade of human impact on each estuary: whereas many modifications have produced major consequences altering the original geomorphology and circulation in the Quequén Grande estuary in the last 100 years, particularly the obstruction of the incoming tidal wave (Perillo et al., 2005), the Quequén Salado mouth dynamics has remained almost unaltered. Since diatom distribution is mainly influenced by the salinity range and sediment type in these estuaries, their morphological differences originated by human modification constitute a key factor in explaining the observed differences in diatom distribution. Hence, diatom assemblages from Río Quequén Salado constituted useful analogues of salinity in low impacted estuaries. Moreover, the data sets from Mar Chiquita, Quequén Grande and Quequén Salado estuaries have been recently used by Hassan et al. (2009) to develop a regional diatom-based salinity transfer function to quantitatively infer past salinity values from fossil diatoms, which will be described below.

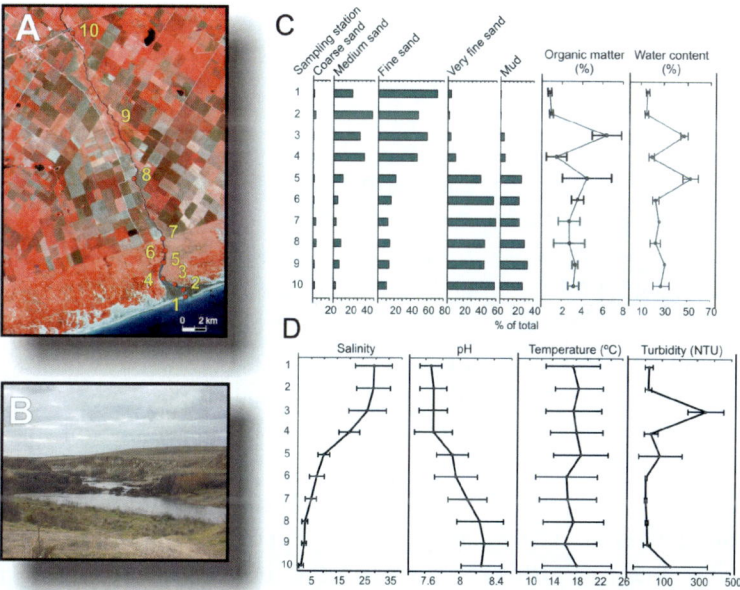

Figure 9. Location of sampling sites (A), view of the estuarine zone (B), and the corresponding sedimentary (C) and water quality (D) parameters at Quequén Salado river (modified after Hassan et al., 2007).

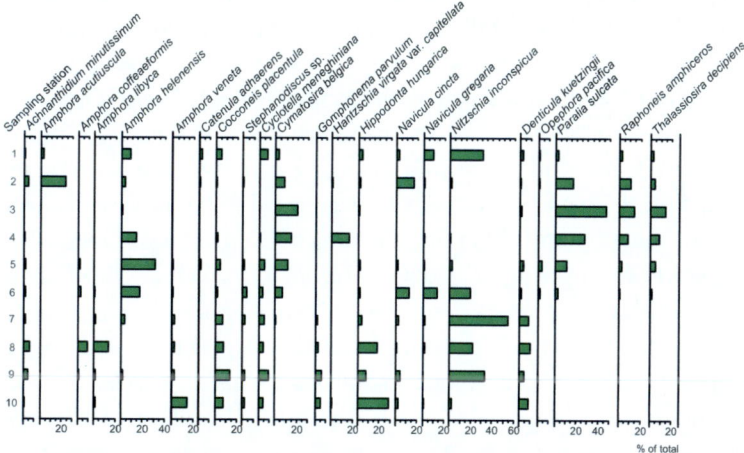

Figure 10. Diatom assemblage composition at Quequén Salado river (modified after Hassan et al., 2009).

## BAHIA BLANCA ESTUARY

Bahía Blanca estuary is a geomorphologicaly complex environment derived from a Late Pleistocene-early Holocene delta complex (Piccolo & Perillo, 1999). It is formed by a series of NW-SE tidal channels separated by extensive intertidal flats, low marshes and islands (Popovich & Marcovecchio, 2008). The northern area is geomorphologicaly dominated by the Main Channel (main navigation channel), while the southern area is dominated by the channels named Bahía Falsa and Bahía Verde, which are the largest within the estuary (Fig. 11). The dominant sedimentology is based on silty clays on the flats and sand in most of the deeper parts of the channels (Piccolo & Perillo, 1999). Mean annual (13°C), summer (21.6°C), and winter (8.5°C) surface water temperatures in the Main Channel are always slightly higher at the head of the estuary (Piccolo et al., 1987), while mean surface salinity increases exponentially from the head to the mid-reaches of the estuary. The water column is vertically homogeneous all throughout the estuary although it may be partially mixed in the inner zone, depending on freshwater runoff conditions. Bahía Blanca estuary includes the largest deepwater harbor system in Argentina, a fact that makes it economically important. This area gathers important urban centers as well as large industrial companies such as a petrochemical industrial park, a thermoelectric plant, fertilizer plants and a commercial duty-free zone on its northern coast (Popovich & Marcovecchio, 2008).

The phytoplankton of Bahía Blanca has been intensively studied during the past decades (Gayoso 1981, 1988, 1998, 1999; Popovich, 2004; Popovich et al., 2008). These studies were mainly focused on the seasonal succession patterns in a fixed station located at the inner part of the Main Channel (Puerto Cuatreros, Fig. 11). The site was characterized by its shallowness and extremely high turbidity (secchi depth <0.5 m), and seasonally changing salinity (22.8 to 41). In these long-term studies, the genus *Thalassiosira* was found to be the most conspicuous component of the phytoplankton in the area. *T. curviseriata* was the most abundant species, followed by *T. anguste-lineata, T. pacifica, T. rotula* and *T. hibernalis. Chaetoceros (Chaetoceros* sp., *C. diadema, C. ceratosporus* var. *brachysetus* and *C. subtilis* var. *abnormis*) was the second most abundant genus. Other important taxa mentioned were *Skeletonema costatum, Ditylum brightwellii, Guinardia delicatula, Asterionellopsis glacialis, Thalassiosira eccentrica, Cyclotella striata, Cerataulina pelagica, Thalassiosira hendeyi, Paralia sulcata,* and *Gyrosigma attenuata.*

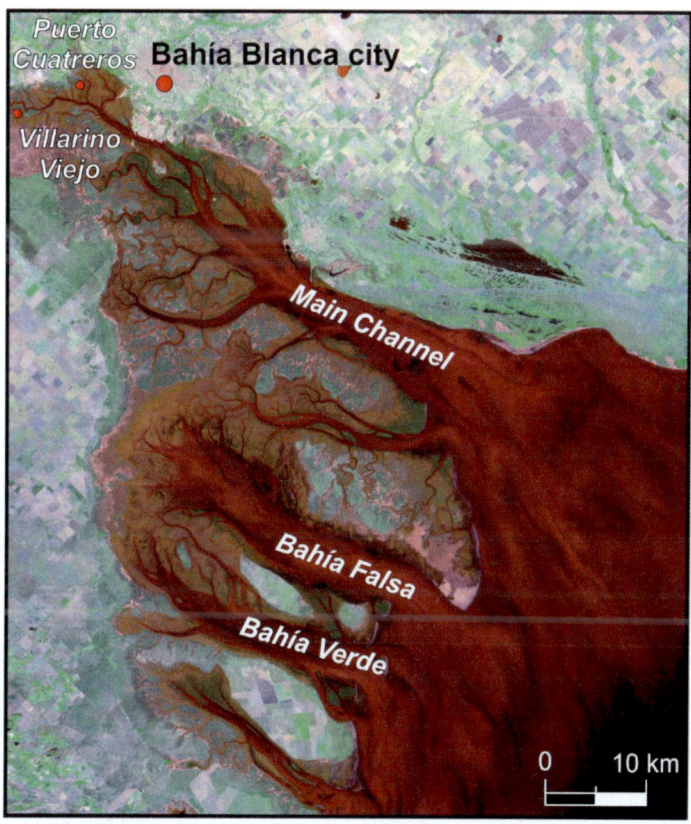

Figure 11. Location map of Bahía Blanca estuary, showing the main channels and the two points studied for diatoms: Puerto Cuatreros and Villarino Viejo.

Literature on the ecology and dynamics of the phytoplankton along the Bahía Blanca estuarine gradient is rather poor, especially towards the outer part of the estuary. In a recent attempt, Popovich & Marcovecchio (2008) studied the spatial and seasonal variation in physical and chemical characteristics and phytoplankton biomass in 9 sites located from the inner to the outer reaches of the estuary. Phytoplankton abundance and nutrient levels (N, P and Si) showed a marked decreasing trend from the head to the mouth of the Bahía Blanca estuary. Mean salinity was relatively constant, from 31.6 in the innermost part to 32.9 in the estuary mouth. Distributional tendencies were exposed in a general qualitative way: the inner and middle zones exhibited a seasonal pattern in diatom assemblages composition: whereas than *Thalassiosira curviseriata, T. angustelineata, T. pacifica, T. rotula, T. hibernalis, T. eccentrica, Chaetoceros ceratosporus, C. diadema, C. debilis* and *Skeletonema costatum* dominated these regions in winter, summer and autumn assemblages were dominated by *Cerataulina pelagica, Guinardia delicatula* and *Cylindroteca closterium*. On the other hand, the occurrence of several marine species such as *Corethron criophilum, Odontella mobiliensis, Coscinodiscus* spp. and *Actinoptychus* spp. at the outer region indicated a higher influence of euhaline offshore waters on this zone of the estuary. Unfortunately, the abundances of individual diatom taxa along the nine sampling sites were not detailed in this contribution, preventing a more precise inference of their autoecological characteristics.

**Table 2. Compositon of the microphytobenthic diatom assemblage and environmental variables in the two sampling sites studied by Parodi and Barria de Cao (2003) at the inner part of Bahía Blanca estuary**

|  | **PUERTO CUATREROS** | **VILLARINO VIEJO** |
|---|---|---|
| **N total (%)** | 0.21 | 0.19 |
| **P extractable (ppm)** | 8.82 | 8.71 |
| **pH** | 8.60 | 8.50 |
| **Salinity** | 35.6 | 34.5 |
| **Temperature (°C)** | 9.20 | 9.40 |
| **Dominant to abundant diatoms** | *Nitzschia* sp.<br>*Pleurosigma fasciola*<br>*Navicula* spp.<br>*Surirella gemma*<br>*Amphripora alata*<br>*Stauroneis* sp.<br>*Scoliopleura* sp.<br>*Cocconeis* sp. | *Nitzschia sigma*<br>*Scoliopleura* sp.<br>*Cocconeis* sp. |
| **Rare diatoms** | *Paralia sulcata?* | *Nitzschia* sp.<br>*Gyrosigma attenuata*<br>*Entomonoeis amphyprora*<br>*Pleurosigma fasciola*<br>*Navicula* spp.<br>*Petrodictyon gemma*<br>*Cylindrotheca closterium*<br>*Amphiprora alata*<br>*Stauroneis* sp.<br>*Paralia sulcata?* |

Sedimentary and microphytobenthic diatoms from Bahía Blanca received much less attention than their phytoplanktonic counterparts. Only one preliminary work (Parodi & Barría de Cao, 2003), which focused on the taxonomic composition of the microalgal mats from the inner part of the estuary (Puerto Cuatreros and Villarino Viejo stations, Fig. 11), was published. Puerto Cuatreros site was closer to the harbor, and hence, more influenced by the suspended sediments and the impact of dredging than Villarino Viejo site. Although both sites exhibited similar values for the physical and chemical parameters measured, the species assemblage of the superficial sediment layers showed important differences (Table 2). Whereas than in Puerto Cuatreros diatoms were the dominant microalgae, in Villarino Viejo mats were dominated by blue-green algae. These differences were attributed to the major disturbance of the former due to the deposition of particles of the fluid mud layer produced by the nearby dredging.

Overall, the analysis of the relatively numerous publications on algae from Bahía Blanca leads to the general conclusion that, although information on single species distribution and environmental preferences does exists, this is presented in a very qualitative and descriptive way that prevents its application in diatom-based paleoenvironmental studies.

## ESTUARIES FROM PATAGONIA AND TIERRA DEL FUEGO

Rivers in the Patagonian region are fed by water originated from the precipitation and/or snow melting on the Andes. They flow across the arid and desert Patagonia region, where practically no tributaries are received. Some of the rivers are considered to be the largest in the country both in valley size and river discharge, such as the Río Colorado, Río Negro and Río Santa Cruz. The climate is semiarid to arid, characterized by strong westerly winds throughout the year (Piccolo & Perillo, 1999). Unfortunately, little is known about the diatoms (and the biota in general) of these estuaries. Only a few contributions on microalgal assemblage composition are available for Río Negro, Bahía San Blas, Río Chubut, Ría de Puerto Deseado and Bahía San Sebastián (see Fig. 1), which are described in the following sections.

*Río Negro estuary:* The Río Negro drains a large basin of 115,800 $km^2$, and its valley is of great importance both for economical and hydrological reasons (Figs. 1 and 12A). River width varies between 500 and 800 m but close to the mouth it has a width of 1 km and flows along a valley of approximately 12 km. Depth ranges from 5 to 10 m. Two banks are found in its mouth, forming an open ebb delta (Piccolo & Perillo, 1999). The river receives the domestic and industrial effluents of the several cities located along their margins, and is regulated by a number of damps and hydroelectric plants located in their tributaries (Pucci et al.,

1996). Only one published study is available for the microalgae of this estuary, which is focused on phytoplanktonic communities (Pucci et al., 1996). In that work, samples were collected from three sampling stations located along the last 30 km of the river, in two seasons (spring and autumn; Fig. 12A). The composition of the assemblages was homogeneous between sites in spring, being *Aulacoseira granulata* and *Asterionella formosa* the dominant diatoms. Sampling in spring was conducted during low tide. Hence, salinity values were low in the three sampling stations (between 0.052 and 0.32). Autumn samples were taken during high tide, and consequently salinity rose up to 26 in the station closer to the mouth, whereas it decreased to values under 0.19 in the other two stations. Accordingly, diatom assemblages were more diverse, and dominated by brackish-freshwater forms in the two inner stations; and by coastal-marine taxa in the outer station (Fig. 12B). Information on nutrients, pH and temperature were also provided (Table 3). Although scarce, the information provided in this work is the only information on modern diatoms from Río Negro. Detailed studies on diatom distribution and variability across the estuarine gradient should be conducted in order to provide useful analogues for paleoenvironmental reconstructions in this estuary.

**Table 3. Measurements of environmental variables at Río Negro estuary (modified after Pucci et al., 1993). Numbers correspond to sampling sites signaled in Figure 12. A: autumn; and S: spring, measurements**

| Station | 1 | | 2 | | 3 | |
|---|---|---|---|---|---|---|
| Season | A | S | A | S | A | S |
| Nitrates (N/L) | 1.83 | 0.22-0.92 | 0.45 | 6.49 | 33.8 | 0.53 |
| Nitrites ($\mu$atgN/L) | 0.43 | 0.05-0.09 | 0.06 | 0.08 | 0.09 | 0.07 |
| Phosphates ($\mu$atgP/L) | 0.8 | 0.13-0.14 | 0.18 | 0.26 | 0.53 | 0.33 |
| Silicates ($\mu$atgSi/L) | 47.9 | 165-181 | 165 | 140 | 147 | 162 |
| Salinity (ppm) | 26.09-26.04 | 0.052-0.31 | 0.02-0.19 | 0.206-0.237 | 0.03 | 0.261 |
| pH | 8.55 | 8.25-8.3 | 7.95-8.1 | 7.8-7.85 | 7.2-7.8 | 8.2-8.25 |

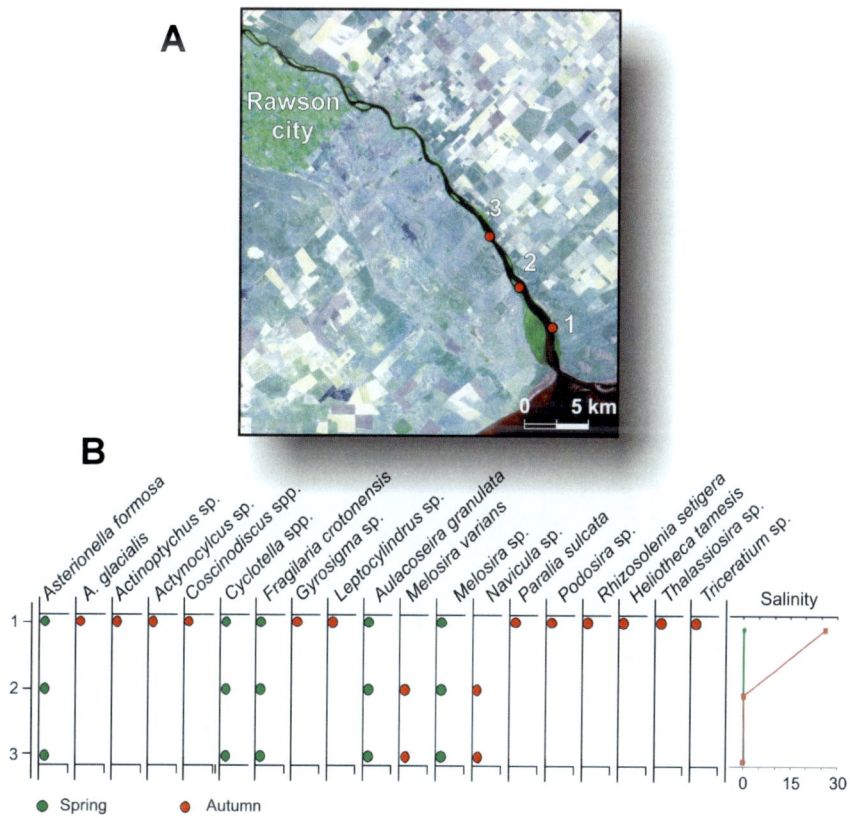

Figure 12. A) Location map of Rio Negro estuary, showing the three sampling points studied by Pucci et al. (1993); B) Distribution of diatom taxa in the sampling sites and the corresponding salinity values in autumn (red) and spring (green), based on data from Pucci et al. (1993).

**Table 4. Diatom assemblage composition of the Jabalí creek samples analyzed by Frenguelli (1938)**

| | Jabalí Creek Samples | | | | | |
|---|---|---|---|---|---|---|
| **Substrate** | Estuarine sediment 1 (mud) | Estuarine sediment 2 (mud) | Beach sediment (sandy mud) | Inside ascidia coenobium (*Julinia* sp.) | Epiphytes under macroalgae (*Stipocaulon* sp. and *Cladophora* sp.) | Epibiotic under bryozoans (*Gemellaria* sp.) |
| **Freshwater diatoms** | *Cocconeis placentula* *C. placentula* var. *lineata* *Coscinodiscus lacustris* *Epithemia adnata* *Pinnularia borealis* | *Epithemia adnata* *Aulacoseira granulata* *Luticola mutica* *Nitzschia frustulum* *Opephora martyi* *Pinnularia borealis* | *Planothidium lanceolatum* *Discostella stelligera* *Epithemia adnata* *Gomphonema gracile* *Hantzschia amphioxys* var. *xerophila* *Aulacoseira granulata* *Luticola mutica* *Navicula peregrina* *Nitzschia frustulum* var. *perpusilla* *Maryana martyi* *Rhopalodia gibba* *R. gibberula* | *Amphora perpusilla* *Epitemia adnata* *Staurosira construens* *Aulacoseira italica* *Nitzschia frustulum* *Rhopalodia gibba* | - | *Encyonema turgidum* *Epithemia adnata* |
| **Brackish diatoms** | *Achnanthes brevipes* var. *intermedia* *Planothidium delicatulum* *Caloneis permagna* *Cyclotella striata* *Diploneis didyma* *D. interrupta* | *Nitzschia clausii* *Rhopalodia musculus* | *Achnanthes brevipes* var. *intermedia* *Planothidium delicathulum* *Gyrosigma balticum* *Nitzschia habirshawii* *N. sigma* var. *rigida* | *Achnanthes brevipes* var. *intermedia* *Planothidium delicatulum* *Gyrosigma spenceri* var. *exilis* *Nitzschia clausii* *Nitzschia sigma* var. *rigida* | - | *Cyclotella baltica* *Gyrosigma balticum* *Bacillaria paradoxa* *B. paradoxa* var. *tropica* |

| | | | | | | |
|---|---|---|---|---|---|---|
| **Marine diatoms** | Gyrosigma balticum<br>G. spenceri var. curvula<br>G. wansbecki<br>Tryblionella compressa<br>Nitzschia sigma<br>Surirella striatula | Amphora granulata<br>Auliscus sculptus<br>Cocconeis scutellum<br>C. scutellum var. parva<br>Paralia sulcata | Paralia sulcata | Amphora angusta<br>Campilosira cymbelliformis<br>Cocconeis scutellum var. ornata<br>C. scutellum var. parva<br>Coscinodiscus excentricus var. minor | Amphora granulata<br>Cocconeis scutellum<br>C. scutellum var. parva<br>Rhoicosphaenia marina<br>Navicula gourdoni<br>N. oceanica<br>N. platyventris | Cocconeis scutellum var. ornata<br>C. scutellum var. minor | Cocconeis scutellum var. ornata<br>C. scutellum var. parva<br>Corethron criophilum |

**Table 5. Diatom assemblage composition of Bahía San Blas samples analyzed by Frenguelli (1938)**

| Bahía San Blas Samples | | |
|---|---|---|
| | Sediment (mud) | Plankton net |
| **Brackish diatoms** | *Achnanthes brevipes*<br>*A. brevipes* var. *intermedia*<br>*Planothidium delicatulum*<br>*Gyrosigma balticum*<br>*Nitzschia clausii*<br>*N. sigma*<br>*N. sigma* var. *sigmatella*<br>*Rhopalodia musculus* | - |
| **Marine diatoms** | *Cocconeis scutellum*<br>*C. scutellum* var. *parva*<br>*Paralia sulcata* | *Biddulphia chilensis*<br>*Odontella mobiliensis*<br>*Lithodesmium undulatum*<br>*Thalassiosira decipiens*<br>*Rhizosolenia imbricata*<br>*Thalassiosira javanica*<br>*Raphoneis amphiceros*<br>*Thalassiosira eccentrica* |

Figure 13. Location map of Bahía San Blas estuary.

*Bahía San Blas estuarine complex:* Only two works on diatoms were performed at Bahía San Blas estuarine complex. The first was conducted by Frenguelli (1938), who analyzed the diatom content of plankton samples and surface sediments in the Jabalí creek (which outflows into the southernmost part of the complex) and in Bahía San Blas harbor (Fig. 13). At Jabalí creek, 6 samples from diverse origins were collected and analyzed (3 of surface sediments, 1 of macroalgae, 1 of ascidians and 1 of bryozoans, Table 4). Samples were dominated by a mixture of brackish and coastal-marine taxa, constituting an estuarine assemblage under a strong tidal influence. In San Blas harbor, one sediment and three plankton samples were analyzed. The assemblage was dominated by marine taxa, although low proportions of brackish diatoms were also recorded (Table 5). Overall, the analyzed samples always showed relatively high proportions of freshwater diatoms (47 taxa were listed), probably transported from the headwaters by the strong winds blowing from the west. Recently, Isla & Espinosa (2005) described the diatom assemblages from a core taken at the Jabalí creek. The top sample of the core (that represents the modern assemblage) was dominated by *Cymatosira belgica* (~20%), *Paralia sulcata* (~10%), *Achnanthes lacus-vulcani* (~10%) and *Planothidium delicatulum* (~20%). Although measurements of salinity (38), pH (7.72) and turbidity (20 NTU) were provided, the application of this punctual datum in paleoenvironmental analyses is limited, since it does not represents the spatial or temporal variability in the composition of diatom assemblages.

*Río Chubut estuary:* The river has a meandering channel, which varies from 70 to 200 m in width, and averages 2 m in depth (Fig. 14A). The river bed shows several sigmoidal bars constituted by medium to coarse sand that divide the river into channels. Waters are rich in silica, and high gradients of $Cl^-$, $Na^+$, $SO_4^{-2}$, $K^+$ and $Mg^{2+}$ are found from the mouth to about 2 km upstream. The river has been dammed at about 120 km from the mouth. It passes through several cities being impacted by agricultural and industrial activities and receiving urban sewages with no or little treatment (Piccolo & Perillo, 1999). A series of works attempted to define the composition of the phytoplankton community in the estuary of the Chubut river (see Appendix I). Sastre et al. (1990) and Villafañe et al. (1991) described the taxa present in the last 9 km of the estuary as a function of salinity. In the inner estuary, salinity was under 3, and the phytoplankton was dominated by *Aulacoseira granulata*, which accounted for more than 80% of the total cells. Although this zone was characterized by a low light penetration (secchi depth: 0.4 m), this did not affect the growth of *A. granulata*, which was able to produce large numbers of individuals under these conditions because it is adapted to low levels of light. Sastre et al. (1994) reported the dominance of this species up to 150 km away from the estuary mouth, where it produced blooms and constituted up to 96% of the total cells. Phytoplankton abundance decreased towards the middle estuary, where salinity ranged between 3 and 30. *A. granulata* was also the most abundant taxon in this zone, although other species of planktonic and benthic diatoms, such as *Biddulphia alternans, B. antediluviana, Gramatophora marina, Triceratium favus, Odontella aurita, Actinoptychus* spp., and *Surirella* spp., were also present. In the outer estuary salinity was higher than 30, and the phytoplanktonic assemblage was dominated by *Odontella aurita*, which accounted for more than 80% of the total cells. Santinelli et al. (1990) analyzed the composition of the community in the mouth of the estuary during two years. Salinity values varied significantly during the tidal cycle from fluvial (0-10) to marine (25-35) conditions. Diatoms were the dominant phytoplanktonic group, being identified 39 taxa, which were grouped by cluster analysis and related to their salinity tolerances (Table 6). One of the groups showed a significant association to low salinity values (group 1, Table 6). The second group (group 2, Table 6) comprised euryhaline taxa, which were distributed all over the estuary. The third group, on the other hand, showed a marked association to the higher salinity values prevailing at the estuary mouth (group 3, Table 6). The defined groups constitute potential analogues useful for paleosalinity reconstructions in Patagonian estuaries.

*Puerto Deseado estuary*: It has a general WSW-ENE orientation, and has an elongated 40 km funnel form (Fig. 14B). Freshwater input comes from the Río Deseado, which used to carry much water during the Pleistocene, but is now reduced to a temporary river. The estuary width varies from 2.5 to 0.4 km, while depth ranges from 5 m in the inner part to 20 m in its mouth. Mean tidal amplitudes range between 4.2 and 2.9 m, and salinity variation is small (<2; Ferrario, 1972; Piccolo & Perillo, 1999). Diatoms from Puerto Deseado estuary were studied by Müller Melcher (1959), who mentioned 12 taxa. In a series of recent contributions (Ferrario, 1972, 1981; Ferrario & Sar, 1984; Ferrario, 1984a,b,c) the list of taxa was expanded to 88 species. For each taxon, a series of taxonomical, ecological and distributional observations were provided. The sampled area was typically marine; salinity ranged between 32 and 34 and pH between 7.5 and 8.4. Nitrates and phosphastes concentrations were of 0.5 mg/l and 0.1 mg/l, respectively. The complete list of diatom taxa mentioned in these works is presented in Appendix II.

## Table 6. Compositon of the phytoplanctonic diatom assemblages along the salinity gradient in the Río Chubut estuary (modified from Santinelli et al., 1990)

| Group | Salinity range | |
|---|---|---|
| | 0-10 ppm | 10-25 ppm |
| Freshwater/ Brackish | *Navicula radiosa*<br>*Navicula* spp.<br>*Cymbella cystula*<br>*Cymbella* spp.<br>*Epithemia sorex*<br>*Rhopalodia gibba*<br>*Cocconeis placentula*<br>*Cocconeis* sp.<br>*Cymatopleura solea*<br>*Surirella* spp.<br>*Asterionella formosa* | |
| Eurihaline | *Paralia sulcata*<br>*Odontella aurita*<br>*Gomphoneis herculeana*<br>*Biddulphia alternans*<br>*Aulacoseira granulata*<br>*Nitzschia* spp.<br>*Thalassiosira* spp.<br>*Gramatophora marina*<br>*Rhabdonema adriaticum*<br>*Biddulphia antediluviana*<br>*Ulnaria ulna*<br>*Synedra* spp.<br>*Melosira varians* | |
| Marine | | *Triceratium favus*<br>*Actinoptychus vulgaris* |

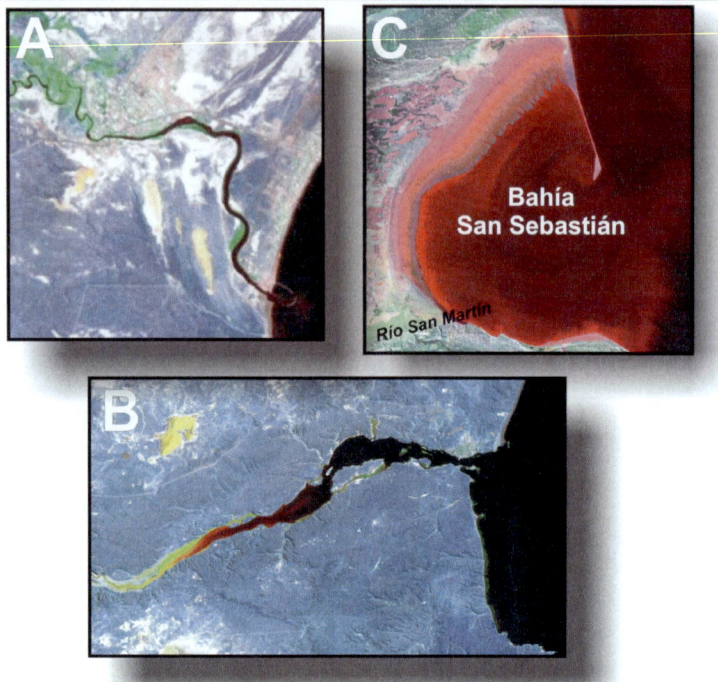

Figure 14. Location maps of A) Río Chubut, B) Puerto Deseado, and C) Bahía San Sebastián estuaries.

**Table 7. Diatom assemblage composition of the Bahía San Sebastián sediment sample analyzed by Frenguelli (1923, 1924)**

| Diatom taxa |
| --- |
| *Actinoptychus senarius* |
| *Thalassiosira eccentrica* |
| *Hyalodiscus radiatus* |
| *Paralia sulcata* |
| *P. sulcata* var. *biseriata* |
| *P. sulcata* var *crenulata*<br>*Psammodictyon panduriforme* var. *parva*<br>*Raphoneis amphiceros*<br>*Surirella striatula*<br>*Surirella tuberosa* var. *costata*<br>*Triceratium scitulum* |

## Table 8. Diatom assemblage composition of the samples from Río Grande estuary, according to Cleve (1900)

| Marine and brackish taxa | Freshwater taxa |
|---|---|
| *Actinoptychus undulatus*<br>*Amphora lineolata*<br>*Biddulphia aurita* | *Amphora pediculus*<br>*Cymbella aspera*<br>*Frustulia rhomboides* |
| *Biddulphia rhombus* | *Hantzschia elongata* |
| *Cocconeis scutellum* var. *genuina* | *Melosira* sp. |
| *Coscinodiscus decipiens* | *Neidium oblique striatum* var. *magellanicum* |
| *Coscinodiscus excentricus* | *Pinnularia borealis* |
| *Coscinodiscus oliverianus* | *Pinnularia commutata* |
| *Entyopyla incurvata* | *Pinnularia elliptica* |
| *Epithemia musculus* | *Pinnularia gibba* |
| *Hantzschia virgata* | *Pinnularia lata* |
| *Hyalodiscus radiates* | *Pinnularia latevittata* |
| *Hyalodiscus scoticus* | *Pinnularia gibba* var. *luculenta* |
| *Melosira nummuloides* | *Pinnularia major* var. *linearis* |
| *Navicula anglica* var. *subsalsa* | *Pinnularia nodosa* |
| *Navicula arenacea* | *Pinnularia stauroptera* |
| *Navicula cincta* | *Pinnularia viridis* |
| *Navicula gregaria* | *Rhoicosphenia curvata* |
| *Navicula pygmaea*<br>*Navicula salinarum*<br>*Navicula subinflata*<br>*Navicula tumida*<br>*Nitzschia apiculata*<br>*Nitzschia constricta* var. *subconstricta*<br>*Nitzschia panduriformis*<br>*Nitzschia sigma*<br>*Paralia sulcata* var. *radiata*<br>*Pleurosigma normanii*<br>*Pleurosigma nubecula* var. *intermedia*<br>*Pleurosigma rigidum*<br>*Podosira maxima*<br>*Rhabdonema arcuatum*<br>*Rhabdonema minutum*<br>*Rhaphoneis amphiceros*<br>*Stauroneis salina*<br>*Surirella gemma*<br>*Surirella striatula*<br>*Triceratium affine* | *Rhopalodia gibba*<br>*Stauroneis phoenicenteron* var. *amphilepta*<br>*Surirella guatemaliensis*<br>*Surirella splendida* var. *tenera* |

*Bahía San Sebastián*: It is a wide bay located in northern Tierra del Fuego, having a semicircular shape partly closed by a long and narrow gravel spit (Fig. 14C). The bay is 55 km long and 40 km wide. The spit has a length of 17 km, and the open mouth is about 20 km wide. Freshwater input into the system is provided by the Río San Martín, which discharges at the southwestern part of the bay. Tidal range is 10 m and wind influence is from the west (Piccolo & Perillo, 1999). Frenguelli (1923, 1924) described the diatom taxa found in a sediment sample collected in San Sebastián bay. Diatoms frustules were scarce. The assemblage was dominated by *Paralia sulcata*, whereas than other ten less frequent taxa were also mentioned (Table 7). No environmental characterization of the sampling point was provided in these studies.

*Río Grande estuary:* The Río Grande flows from west to east, receiving tributaries from the south and the north. Before discharging into the Atlantic Ocean, the river makes a long bend to the south around gravel beach barriers on which the Río Grande is built (Fig. 1). The inlet is therefore constrained by gravel spits that have a significant morphologic variability. The mean tidal range in Río Grande outer estuary is 4.16 m (Isla & Bujalesky, 2004). The only work on diatoms from the Río Grande estuary was carried out by Cleve (1900), who analyzed a series of samples of the estuarine area and provided lists of marine-brackish (38 taxa) and freshwater (22 taxa, Table 8) forms. This work is taxonomic and does not include environmental information.

*Chapter 3*

# HOW CAN RESEARCHERS IMPROVE THE QUALITY OF DIATOM-BASED PALEOENVIRONMENTAL INFERENCES IN COASTAL SETTINGS?

The application of diatom autoecology to paleoenvironmental reconstructions has a long history in the Argentinean coast. Pioneer studies were conducted by Frenguelli (1924, 1925, 1945), who described diatom assemblages present in Holocene successions outcropping in estuaries along the Pampean coast. Besides some of the major estuaries described in the previous section (Río Quequén Grande, Río Quequén Salado and Bahía Blanca), many small streams that flow into the Río de la Plata or the Atlantic coast were included. A total of 276 diatom taxa were listed, from which only 11 species were present in high proportions and formed the dominant assemblage. These were: *Campylodiscus clypeus, Cocconeis placentula, Denticula valida, Diploneis argentina, Hyalodiscus subtilis, Nitzschia vitrea, Rhopalodia gibberula, R. argentina, Surirella striatula* and *Synedra platensis*. Overall, diatom assemblages indicated the presence of environments under marine influence that evolved to brackish/freshwater continental conditions, and ended in swamps which finally got dry as a consequence of the climatic aridization.

During the last 20 years, the paleoenvironmental evolution of the southern Pampas coast and its relationship to the Holocene sea-level fluctuations have been inferred from the detailed study of sedimentary successions originated by the infilling of estuarine sediments. The analyses were based on the diatoms autoecological classifications of salinity and life form taken from De Wolf (1982), Vos and De Wolf (1988, 1993), and Denys (1991/1992), and allowed to infer the presence of sedimentary environments characterized by different salinities and depths. Between ca. 6700 and 3900 $^{14}$C yr BP, the marine influence related to the sea-level high stand was the dominant forcing on paleosalinity trends, occurring at different times and magnitudes according to the characteristics of each basin (Isla et al., 1986; Espinosa, 1998). In the area of Arroyo La Ballenera (see Fig. 1) an estuarine lagoon with small or no tidal range was inferred for the interval between ca. 6200 and 4800 $^{14}$C yr BP, whereas in Arroyo Las Brusquitas (Fig. 1) estuarine conditions lasted up to ca. 3900 $^{14}$C yr BP (Espinosa et al., 2003). In Punta Hermengo area (Fig. 1), a tidal channel infilling was inferred at ca. 6700 $^{14}$C yr BP (Espinosa, 2001). In Río Quequén Grande, the maximum saline influence was detected between ca. 7100 and 5350 $^{14}$C yr BP at 2 km from the river mouth in relation to the development of an estuarine lagoon (Espinosa, 1988, 1998). This marine influence was not recorded in synchronic deposits outcropping 32 km upstream from the previous site (Zárate et al., 1998). In Río Quequén Salado, the analysis of diatom assemblages from two sequences outcropping at 20 and 30 km from the estuary mouth revealed the presence of fluvial-lacustrine environments during the late Pleistocene, followed by alluvial plains with a pulse of marine influence, which finally evolved to lacustrine environments that became brackish and shallower towards the early Holocene (Schillizzi et al., 2006). In the sector of the Pampas coast located between Río Quequén Salado and Bahía Blanca, diatom analyses allowed to infer the development of estuarine lagoon environments during the middle Holocene (between ca. 6500 and 6900 years BP). These estuarine lagoons were transgressed by the sea towards the late Holocene (ca. 5300-4800 $^{14}$C years BP; Gutiérrez Téllez & Schillizzi, 2002; Aramayo et al., 2005).

In contrast to the abundant information available on coastal Holocene diatoms from the Pampean region, data from Patagonia are scarce and studies initiated only recently (see Espinosa, 2008). Isla & Espinosa (2005) analyzed the evolution of southern Bahía San Blas during the late Holocene. The dominance of marine and marine/brackish diatom assemblages in a sediment core obtained in the Jabalí Creek suggested that the zone maintained a hypersaline regime during the last 4700 years. Escandell et al. (2009) analyzed the diatom assemblages from a core obtained 9 km upstream from the Río Negro mouth, in order to reconstruct the late Holocene paleoenvironmental evolution of the estuary. In this contribution, both European ecological codes as well as modern information provided by Hassan (2008) for pampean estuaries were applied. The sequence, which comprised the interval between 2027±34 $^{14}$C years BP and the present, recorded the evolution of a shallow vegetated brackish-freshwater environment at the bottom, which evolved towards a tidal channel that declined gradually in depth and salinity to the middle, to finally end in a marsh influenced by tides and floods towards the top of the sequence.

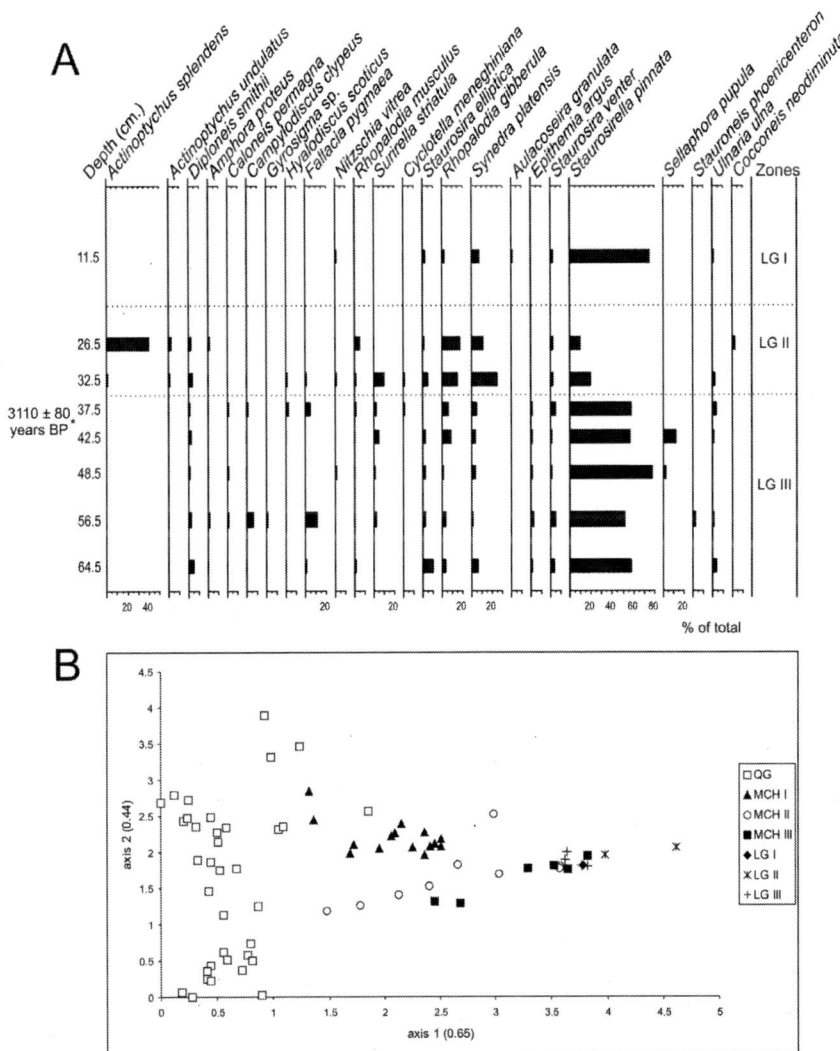

Figure 15. A) Relative frequencies of diatom taxa in the Las Gallinas Creek sequence; B) DCA of combined surface (QG: Quequén Grande, MCHI, MCHII, MCHIII: Mar Chiquita tidal inlet, inner lagoon and headwaters, respectively), and fossil diatom samples (LG: Las Gallinas Creek). Diatom zones were delimited through cluster analyses (reproduced from Hassan et al., 2006; with permission).

The first attempt to apply data on local modern diatom distribution to paleoenvironmental reconstruction in estuarine settings of Argentina was carried out by Hassan et al. (2006). In this work, modern data from Mar Chiquita and Quequén Grande estuaries were compared with fossil data obtained from a late Holocene sequence outcropping at the headwaters of the Mar Chiquita coastal lagoon (Arroyo Las Gallinas) through the application of semi-quantitative techniques (DCA ordination). The sequence had been previously studied through autoecological techniques (Espinosa, 1994). All diatom assemblages were dominated by oligohalobous indifferent taxa (*Staurosirella pinnata* and *Staurosira venter*), accompanied by some oligohalobous halophilous and mesohalobous taxa (such as *Staurosira elliptica*, *Fallacia pygmaea* and *Campylodiscus clypeus*), except for a level located near the middle of the sequence that was dominated by the polyhalobous *Actinoptychus splendens*, the mesohalobous *Rhopalodia musculus* and the oligohalobous halophilous *R. gibberula* (Fig. 15). DCA ordination of modern and fossil samples showed that, except for this level, fossil diatoms from Arroyo Las Gallinas were analogue to modern diatom assemblages living today in the inner lagoon of Mar Chiquita (sites 14 and 15 in Fig. 4A), representing a shallow brackish/freshwater environment, with low salinity fluctuations (1-9) and no tidal influence. Espinosa (1994) proposed tidal channel conditions for the basal levels of Las Gallinas sequence, based on the presence of silty clays and the dominance of tychoplankton. Espinosa (1998) reinterpreted Las Gallinas paleoenvironments as shallow brackish environments with low tidal influence and significant freshwater inflow. On the basis of modern data analysis, Hassan et al. (2006) discarded the tidal influence, since there was no similarity between fossil levels and modern assemblages from Mar Chiquita tidal zone.

In an attempt to increase the accuracy of coastal paleoenvironmental reconstructions in southern Pampas, Hassan et al. (2009) conducted the first quantitative reconstruction of past environmental parameters in estuarine environments of Argentina. In this contribution, the modern data sets provided by Hassan et al. (2006, 2007) for Mar Chiquita coastal lagoon, Río Quequén Grande and Río Quequén Salado were integrated to construct a diatom-based salinity calibration model, based on Weighted Averaging Partial Least Squares techniques (WA-PLS, ter Braak & Juggins, 1993). WA-PLS, together with its simpler version Weighted Averaging (WA), constitute the most robust and simple regression techniques available for quantitative reconstructions based on unimodal distributions (ter Braak et al., 1993; Birks, 1995). In a first step, the relationship between the 48 dominant diatom taxa and salinity was evaluated, and optima and tolerances for each taxon were calculated (Fig. 16).

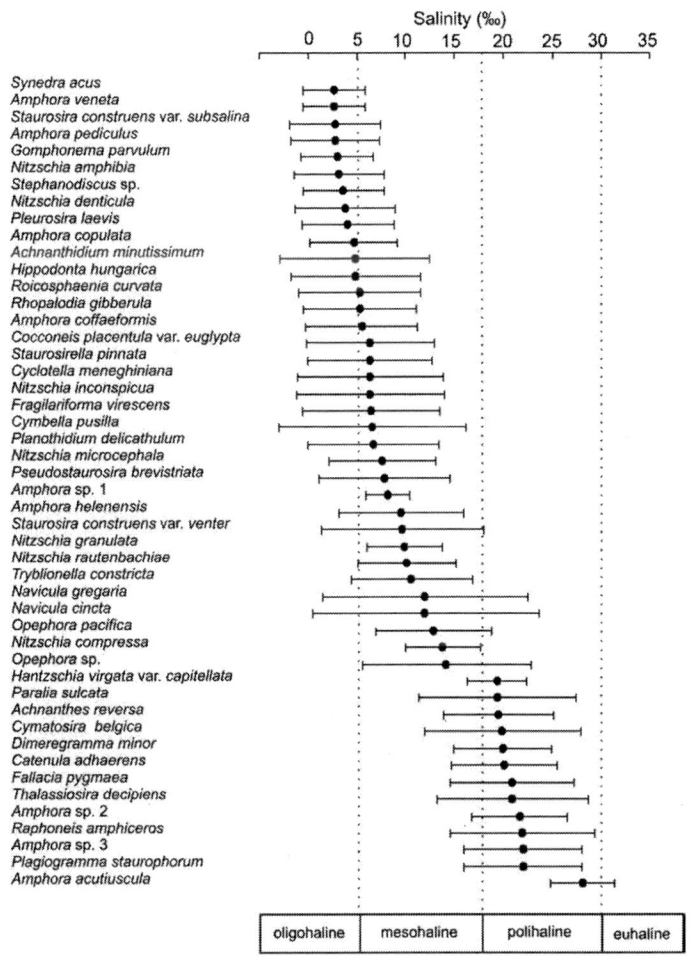

Figure 16. Plot of the salinity optima and tolerances of diatom taxa calculated from Mar Chiquita, Quequén Salado and Quequén Grande datasets. Salinity classification follows Day, 1981 (reproduced from Hassan et al., 2009; with permission).

According to their salinity optima, diatom taxa were divided into three groups: a freshwater group, with salinity optima in oligohaline waters (up to 5); a brackish group, distributed in mesohaline waters (5–18) and a polyhalobous group, restricted to polihaline waters (18–30; Day, 1981). According to their salinity tolerances, most taxa can be regarded as markedly euryhaline (Denys, 1991/1992), since they tolerated salinity changes between 2.3 and 11.6. Taxa located at both ends of the diagram (freshwater and polyhalobous taxa) showed the narrowest tolerance ranges, whereas mesohaline taxa showed the widest ones

(Fig. 16). The salinity transfer function constructed on the basis of this data set showed a good performance, with an error of 4.42, comparable to the obtained in salt marshes from North America (Sherrod, 1999).

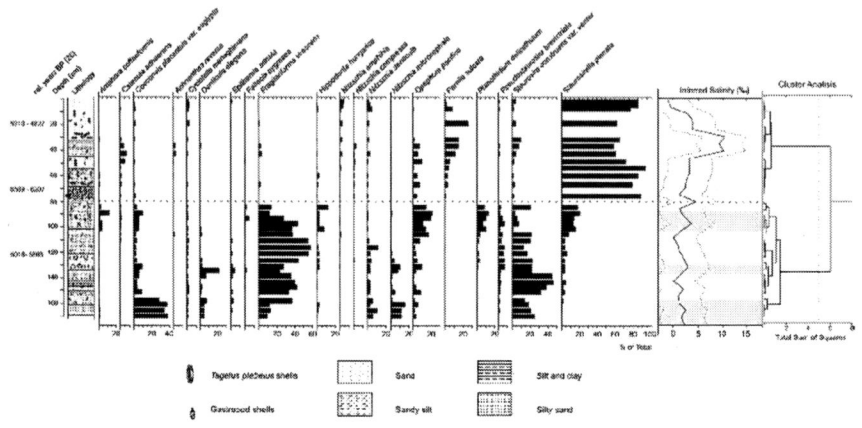

Figure 17. Lithology, relative frequency diagram of diatom composition and inferred salinity values at Puente Taraborelli profile. Grey shadows indicate salinity values inferred from samples that lack good analogues in the training set (reproduced from Hassan et al., 2009; with permission).

The modern data set was applied to the paleoenvironmental reconstruction of a sedimentary sequence outcropping at the left margin of the Río Quequén Grande (Puente Taraborelli section, site 13 in Fig. 7A), 12 km upstream from the estuary mouth. Diatom assemblages of the basal and medium sections of the sequence (0.8–1.8 m in depth) were dominated by *Fragilariforma virescens*, *Staurosira venter*, *Cocconeis placentula* var. *euglypta*, *Denticula kuetzingii*, *Nitzschia inconspicua* and *Planothidium delicatulum*. Samples from the top of the sequence (0–0.8 m in depth) were dominated by *Staurosirella pinnata*, accompanied by *Staurosira venter*, *Catenula adhaerens* and *Paralia sulcata* (Fig. 17). In a semi-quantitative approach, modern and fossil samples were ordered in a two dimensional space through DCA (Fig. 18). Results of DCA ordination showed that Holocene diatom assemblages were more similar to the modern diatom assemblages from Mar Chiquita than those living today at Quequén Grande river, suggesting the presence of an estuarine lagoon rather than an estuary of lotic characteristics. The application of the transfer function to the fossil diatom assemblages allowed the quantitative reconstruction of Holocene salinity fluctuations (Fig. 17). Maximum salinity values, estimated at about 13, were detected between ca. 7500±90 and 6040±90 $^{14}$C yr BP. Therefore, the integration of these results to those obtained in previous works (Espinosa, 1998) suggested that the marine influence in Quequén Grande occurred since ca. 7500 $^{14}$C yr BP, extending up to 12 km from the present coastline through ca. 7000 $^{14}$C yr BP, in relation to the development of an estuarine lagoon of large dimensions. In contrast to these quantitative results, the application of the autoecological classifications (*sensu* Vos & De Wolf, 1993) only allowed the recognition of two main sedimentary environments within broad salinity compartments: diatom assemblages indicated a brackish/freshwater environment of continental characteristics in the basal and medium sections of the sequence, and a marine/brackish environment subjected to small tidal range towards the top of the sequence (Fig. 19).

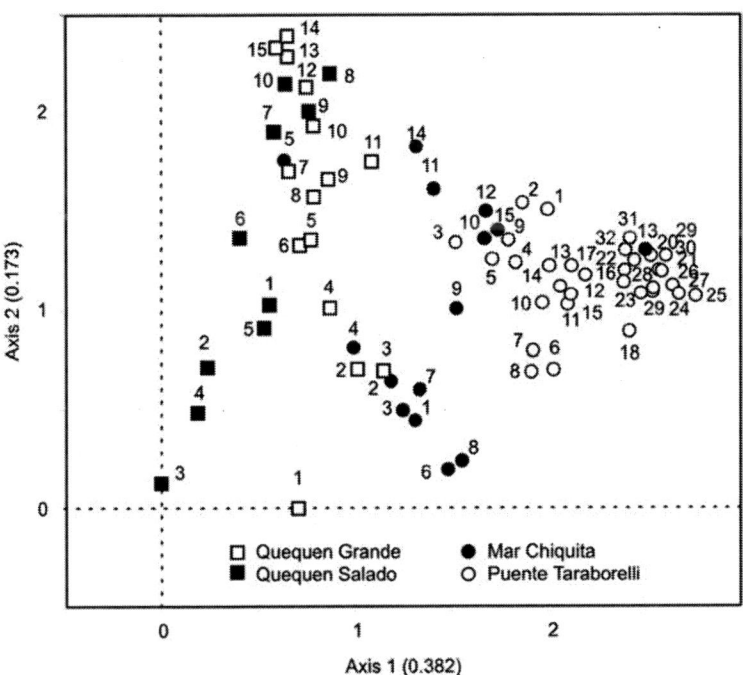

Figure 18. Results of DCA ordination of modern diatom samples from Mar Chiquita, Quequén Grande and Quequén Salado estuaries, and fossil diatom samples from Puente Taraborelli sequence. MCH, QG and QS numbers correspond to sampling sites showed in Figs. 4, 7 and 9 (reproduced from Hassan et al., 2009; with permission).

It becomes clear that to the general characterization of sedimentary environments provided by the autoecological techniques widely applied in the region, the transfer function approach adds a method to map both temporal and spatial variations in paleosalinity values. The wide salinity tolerances of estuarine diatoms found in the three studied estuaries restricts the accuracy of the paleoenvironmental reconstructions based on their autoecology, even when autoecological data are obtained from local environments. A clear example of this is the euryhaline species *Staurosirella pinnata*, which dominates Holocene sucessions of both estuarine (e.g. Hassan et al., 2009) and freshwater (e.g. Espinosa, 1994) origin, limiting the paleoenvironmental inferences that can be done from the assemblage. This limitation is strongly linked to the impossibility of classifying individual taxa into narrow salinity classes, problem that is saved by applying synecological techniques, since they are based on a weighted average of the optima and tolerances of all taxa present in a fossil sample. Accordingly, researchers can improve the quality of diatom-based paleoenvironmental reconstructions by incorporating quantitative approaches to their projects. Furthermore, it would be useful to generate modern data sets that allowed a semi-quantitative analysis of fossil data by detecting and identifying modern environments that could possibly be analogue to the ones that developed during the Holocene. Even when this approach does not provide quantitative estimates of past environmental variables it supplies a useful tool to assess the paleoenvironmental significance of fossil diatom assemblages dominated by taxa with broad salinity tolerances.

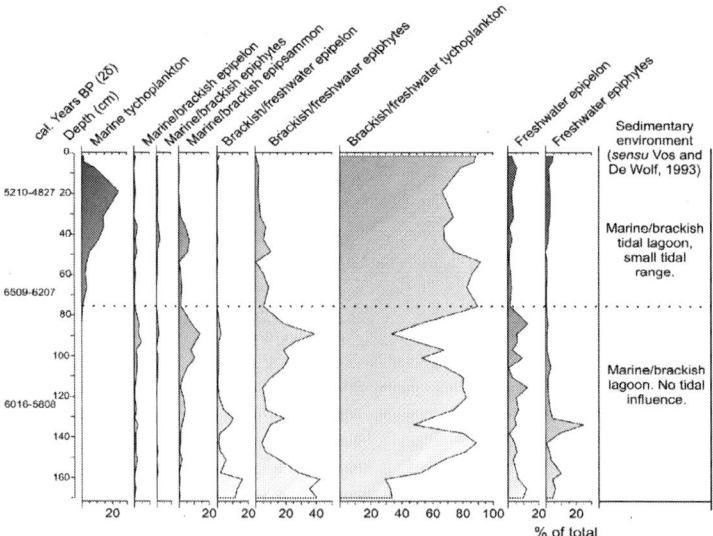

Figure 19. Relative frequency diagram of diatom ecological groups and their environmental significance according to Vos and De Wolf, 1993 (modified from Hassan, 2008).

*Chapter 4*

# CONCLUSIONS

The bibliographic analysis carried out in the previous sections evidences that information on modern diatoms from Argentinean estuaries is very scarce and fragmentary, a fact that clearly contrasts with the abundance, magnitude and economic importance of these environments in the region. In general, most of the reviewed works focused on diatom assemblages from the more densely inhabited Pampean coast, whereas estuaries from Patagonia, in some cases less accessible, received less attention. From the analysis of Appendix II, a general tendency of increasing diatom richness towards the south can be recognized: the highest number of taxa was mentioned for Río de la Plata (n= 356), whereas the lowest values were recorded in Bahía San Blas (n=15) and Río Negro (n=19). Although some geographical component could be invoked to explain this apparent tendency, many problems with the dataset pose serious limitations to the formulation of general biogeographical conclusions. Works contrasted significantly in sampling strategy and intensity, as well as in the ecological compartment studied. For example, while some studies only dealt with one sample (e.g. Frenguelli 1923, 1924), others included more than 50 samples (e.g. Licursi et al., 2006; Hassan et al., 2009). Moreover, studies focused either on phytoplanktonic (e.g. Licursi et al., 2006; Pucci et al., 2006), sedimentary (e.g. Parodi & Barría de Cao, 2003; Hassan et al., 2009) or epiphytic (e.g. Bauer et al., 2007) assemblages. It is evident that the more samples are analyzed, the more different and numerous taxa that can be found. It is also obvious that different habitats contain different diatom floras. Hence, the methodological inconsistency underlying the data set does not allow performing comparisons on diatom diversity and biogeography aspects.

Many of the reviewed works supplied some kind of environmental information (particularly on salinity), which is one of the main requisites to apply the information on diatom assemblage composition to reconstruct past environments. However, studies differed significantly in the quality of the datasets provided. The most complete and useful works were those in which the research was guided by autoecological or paleoecological objectives. These detailed distributional studies were carried out in the Río de la Plata (Licursi et al., 2006), Mar Chiquita, Quequén Grande and Quequén Salado (Hassan et al., 2009) estuaries. They provided information not only on single diatom taxa distribution but also on environmental parameters along the estuarine gradient, allowing extracting either autoecological or quantitative data applicable to the fossil record.

Unfortunately, a great number of the available works included only punctual samplings, restricted either in time or in space, which do not reflect the high variability of diatom assemblages. The most significant example was the Bahía Blanca estuary, where a relatively large number of detailed studies on diatom seasonality were conducted in the last decades, but mostly restricted to one single site located in the inner estuary (Puerto Cuatreros). In other cases, studies covered neither spatial nor temporal variability on diatom assemblage composition, since samples were taken only in one (e.g. Bahía San Blas; Isla and Espinosa, 2005) or two (e.g. Río Negro; Pucci et al., 1996) moments of the year. This is an essential issue when working in estuarine environments, which are subjected to significant environmental fluctuations to which diatoms must become adapted. Consequently, it is not possible to assess the environmental preferences of the diatom taxa present in a sample if the whole range of variability in estuarine conditions has not been covered by the sampling strategy. Hence, the report of the presence of a taxon at a given salinity value in a sole sampling point, as provided in many of the reviewed studies, constitutes only an anecdotal data of restricted applicability for paleoenvironmental reconstructions.

There were also some works which, although based on detailed and well-planned sampling strategies, did not present the results in an accessible way. Examples of these are found in the Río de la Plata (Gómez et al., 2009) and Bahía Blanca (Popovich & Marcoveccio, 2009) estuaries, where although diatom assemblages from a set of sampling sites distributed along the estuarine gradient were studied and environmental data presented, the frequencies or abundances of taxa in each site were not provided. This omission prevented the linking of each taxon to the values of the environmental parameters at which they were found, information that would have resulted very useful for paleoenvironmental reconstructions. In other words, there is a large amount of information but it is unavailable to the reader. This constitutes one of the most surprising findings of the present review, since evidences a lack of contact between ecologists and paleoecologists that may lead to an unnecessarily doubling of research efforts.

If the problems listed above are taken into consideration, the information summarized in Appendix II can be reliably applied to paleoenvironmental reconstructions. However, it is necessary to be aware of the fact that the salinity information listed represents the ranges at which each taxon was found in Argentinean estuaries, and not its optimal and tolerance (excepting for a few exceptions in which these parameters were statistically calculated). Likewise, the type of sample (plankton, sediment, or vegetation) at which each taxon was recorded in each estuary does not necessarily coincide with the habitat of that species. In some cases, taphonomic processes can resuspend benthic diatoms and incorporate them into the water column, while in others plankton forms can be found deposited in surface sediments (Juggins, 1992). Examples of these are presence of planktonic taxa (such as *Actinoptychus splendens* or *Actinocyclus octonarius*) in sediments of Mar Chiquita coastal lagoon, as well as the finding of non-planktonic species (such as *Cocconeis placentula* and *Gomphonema parvulum*) in plankton samples from the Río de la Plata estuary.

Finally, it should be noted that progress in Holocene estuarine diatom paleoecology in Argentina will greatly depend on further study of all aspects of modern diatom ecology and distribution, as well as of the taphonomic processes that alter dead diatom frustules before and during its deposition. In that way, there are many issues that need to be investigated, such as the nature and extent of the taphonomic biases suffered by plankton assemblages; the detailed distribution patterns of diatom assemblages along the environmental gradient of most of the Argentinean estuaries; the single taxa optima and tolerances for key environmental factors, and the biogeographical distributional patterns. The observation of modern environments would not only allow a better knowledge of the environmental significance of fossil assemblages, but also to construct new hypothesis to guide future investigations in paleoecological research.

# ACKNOWLEDGMENTS

I am very grateful to Frank Columbus for inviting me to participate of this publication. This work would not have been possible without the help, patience, and critical comments of Claudio De Francesco. I would also like to acknowledge Silvia De Marco, Hugo Freije, Gerardo Perillo, and Viviana Sastre for gently providing copies of their works. José María Guerrero, Patricio Rivera, Fabricio Idoeta, and Eleonor Tietze helped in the compilation of the bibliography. Felipe García-Rodríguez and Marcela Espinosa made helpful commentaries and suggestions on the manuscript. Landsat 5 and 7 satellite images were freely supplied by the Comisión Nacional de Energía Atómica (CONAE) through its webpage (http://www.conae.gov.ar). Financial support for this work was provided by University of Mar del Plata (EXA 457/09). G.S.H. is a member of the Scientific Research Career of the Consejo Nacional de Investigaciones Científicas y Técnicas (CONICET).

# REFERENCES

Acha, M. E., Mianzan, H., Guerrero, R., Carreto, J., Giberto, D., Montoya, N. & Carignan, M. (2008). An overview of physical and ecological processes in the Rio de la Plata Estuary. *Palaeogeography, Palaeoclimatology, Palaeoecology, 275,* 77–91.

Admiraal, W. (1977a). Influence of light and temperature on the growth rate of estuarine benthic diatoms in culture. *Marine Biology, 39,* 1-9.

Admiraal, W. (1977b). Salinity tolerance of benthic estuarine diatoms as tested with a rapid polarographic measurement of photosynthesis. *Marine Biology, 39,* 11-19.

Admiraal, W. (1977c). Influence of various concentrations of orthophosphate on the division rate of an estuarine benthic diatom, *Navicula arenaria*, in culture. *Marine Biology, 42,* 1-8.

Admiraal, W. (1977d). Tolerance of benthic diatoms to high concentrations of ammonia, nitrite ion, nitrate ion and orthophosphate. *Marine Biology, 43,* 307-315.

Admiraal, W. (1984). The ecology of estuarine sediment inhabiting diatoms. In F. E. Round, & D. J. Chapman (Eds.), *Progress in Phycological Research* 3 (pp. 269-322). Bristol: Biopress Limited.

Admiraal, W. & Peletier, H. (1980). Distribution of diatom species on an estuarine mud flat and experimental analysis of the selective effect of stress. *Journal of Experimental Marine Biology and Ecology, 46,* 157-175.

Admiraal, W., Peletier, H. & Zomer, H. (1982). Observations and experiments on the population dynamics of epipelic diatoms from an estuarine mudflat. *Estuarine Coastal and Shelf Science, 14,* 471-487.

Alin, S. R. & Cohen, A. S. (2004). The live, the dead, and the very dead: taphonomic calibration of the recent record of paleoecological change in Lake Tanganyika, East Africa. *Paleobiology, 30,* 44-81.

Ampsoker, M. C. & McIntire, C. D. (1978). Distribution of intertidal diatoms associated with sediments in Yanquina Estuary, Oregon. *Journal of Phycology, 14,* 387-395.

Aramayo, S. A., Gutierrez-Tellez, B. & Schillizzi, R. A. (2005). Sedimentologic and paleontologic study of the southeast coast of Buenos Aires province, Argentina: a late Pleistocene-Holocene paleoenvironmental reconstruction. *Journal of South American Earth Sciences, 20,* 65-71.

Ayestaran, M.G. & Sastre, A.V. (1995). Diatomeas del curso inferior del Río Chubut (Patagonia Argentina). Pennales I: Naviculaceae. *Boletin de la Sociedad Argentina de Botánica, 31,* 57-68.

Austen, I., Andersen, T. J. & Edelvang, K. (1999). The influence of benthic diatoms and invertebrates on the erodibility of an intertidal mudflat, the Danish Wadden Sea. *Estuarine, Coastal and Shelf Science, 49,* 99-111.

Balech, E. (1976). Fitoplancton de la campaña convergencia 1973. *Physis, 35,* 47-58.

Balech, E. (1978). Microplancton de la campaña productividad IV. *Revista del Museo Argentino de Ciencias Naturales, Hidrobiología 5,* 137-201.

Bao, R., Alonso, A., Delgado, C. & Pagés, J. L. (2007). Identification of the main driving mechanisms in the evolution of a small coastal wetland (Traba, Galicia, NW Spain) since its origin 5700 cal yr BP. *Palaeogeography, Palaeoclimatology, Palaeoecology, 247,* 296–312.

Battarbee, R. W., Charles, D. F., Dixit, S. S. & Renberg, I. (1999). Diatoms as indicators of surface water acidity. In E. F. Stoermer, & J. P. Smol (Eds.), *The diatoms: applications for the environmental and Earth Sciences* (pp. 85-129). London: Cambridge University Press.

Bauer, D. E., Gómez, N. & Hualde, P. R. (2007). Biofilms coating *Schoenoplectus californicus* as indicators of water quality in the Río de la Plata Estuary (Argentina). *Environmental Monitoring Assessment, 133,* 309-320.

Bayssé, C., Elgue, J.C., Burone, F. & Parietti, M. (1986). Campaña de invierno 1983. II Fitoplancton. *Publicaciones de la Comisión Técnica Mixta del Frente Marítimo, 1,* 218-229.

Behrensmeyer, A. K., Kidwell, S. M. & Gastaldo, R. A. (2000). Taphonomy and paleobiology. *Paleobiology (supplement), 26,* 103-147.

Bennett, J. R., Bianchi, T. S. & Means, J. C. (2000). The effects of PAH contamination and grazing on the abundance and composition of microphytobenthos in salt marsh sediments (Pass Fourchon, LA, USA). II. The use of plant pigments as biomarkers. *Estuarine, Coastal and Shelf Science, 50,* 425-439.

Bergamasco, A., De Nat, L., Flindt, M. R. & Amos, C. L. (2003). Interactions and feedbacks among phytobenthos, hydrodynamics, nutrient cycling and sediment transport in estuarine ecosystems. *Continental Shelf Research, 23,* 1715-1741.

Bianchi, T. S. & Rice, D. L. (1988). Feeding ecology of *Leitoscoloplos fragilis.* II. Effects of worm density on benthic diatom production. *Marine Biology, 99,* 123-131.

Birks, H. J. B. (1995). Quantitative palaeoenvironmental reconstructions. In D. Maddy, & J. S. Brew (Eds.), *Statistical modelling of Quaternary science data* (pp. 161-254). Cambridge: Quaternary Science Association.

Birks, H. J. B. & Birks, H. H. (1980). *Quaternary palaeoecology.* Baltimore: University Park Press.

Campeau, S., Pienitz, R. & Héquette, A. (1999). Diatoms as quantitative paleodepth indicators in coastal areas of the southeastern Beaufort Sea, Arctic Ocean. *Palaeogeography, Palaeoclimatology, Palaeoecology, 146,* 67-97.

Calliari, D., Gómez, M. & Gómez, N. (2005). Biomass and composition of the phytoplankton in the Río de la Plata: large-scale distribution and relationship with environmental variables during a spring cruise. *Continental Shelf Research, 25,* 197–210.

Calliari, D., Brugnoli, E., Ferrari, G. & Vizziano, D. (2009). Phytoplankton distribution and production along a wide environmental gradient in the South-West Atlantic off Uruguay. *Hydrobiologia, 620,* 47-61.

Carbonell, J.J. (1935). Some micrographic observations of the waters of the River Plate. *Verhandl. Internat. Vereinig. F. theor. u . angewandle Limnologie, 7.*

Carbonell, J.J. & Pascual, A. (1925). Una *Melosira* nueva para el Río de la Plata. *Physis, 8,* 106-107.

CARP-SIHN-SOHMA (Com. Adm. Río de la Plata –Serv. Hidrog. Naval Argentina –Serv. Oceanog. Hidrol. Meteorol. Armada Uruguay). (1989). *Estudios para la evaluación de la contaminación en el Río de la Plata.* Buenos Aires-Montevideo, 422 pp.

Carreto, J. I., Montoya, N. G., Benavides, H. R., Guerrero, R. & Carignan, M. O. (2003). Characterization of spring phytoplankton communities in the Río de La Plata maritime front using pigment signatures and cell microscopy. *Marine Biology, 143,* 1013-1027.

Carreto, J. I., Montoya, N., Akselman, R., Carignan, M. O., Silva, R. I. & Cucchi Colleoni, D. A. (2008). Algal pigment patterns and phytoplankton assemblages in different water masses of the Río de la Plata maritime front. *Continental Shelf Research, 28,* 1589-1606.

Cervetto, G., Mesones, C. & Calliari, D. (1987). Phytoplankton biomass and its relationship to environmental variables in a disturbed coastal area of the Río de la Plata, Uruguay, before the new sewage collector system. *Atlântica, 24,* 45-54.

Colijn, F., Admiraal, W., Baretta, J. W. & Ruardij, P. (1987). Primary production in a turbid estuary, the Ems-Dollard: field and model studies. *Continental Shelf Research, 7,* 1405–1409.

Cleve, P. T. (1894/1895). Synopsis of the naviculoid diatoms. *Kgl. Sven. Vet. Akad. Handl 26/27,* 1–219.

Cleve, P.T. (1900). Report on the diatoms of the Magellan territories. *Svenska Expeditionen till Magellansläderna, 3,* 273-283.

Cooper, S. R. (1999). Estuarine paleoenvironmental reconstructions using diatoms. In E. F. Stoermer, & J. P. Smol (Eds.), *The diatoms: Applications for the environmental and Earth Sciences* (pp. 352-373). London: Cambridge University Press.

Cordini, J.M. (1939). El seston del Río de la Plata y su contenido diatómico. *Revista del Centro de Estudios de Doctorado de Ciencias Naturales, 2,* 157-169.

Day, J. H. (1981). *Estuarine ecology, with particular reference to southern Africa.* Rotterdam: A.A. Balkema.

Day, J. W., Hall, C. A. S., Kemp, W. M. & Yáñez-Arancibia, A. (1989). *Estuarine Ecology.* John Wiley & Sons, Inc., (eds). New York, Chichester, Brisbane, Toronto, Singapore, 558 pp.

De Francesco, C. G. & Isla, F. I. (2003). Distribution and abundance of hydrobiid snails in a mixed estuary and a coastal lagoon, Argentina, *Estuaries, 26,* 790-797.

De Marco. S. G. (2002). *Características hidrológicas y bioópticas de las aguas de la laguna costera Mar Chiquita y su relación con el fitoplancton.* Doctoral Thesis, Universidad Nacional de Mar del Plata.

De Marco, S. G., Beltrame, M. O., Freije, R. H. & Marccovecchio, J. E. (2005). Phytoplankton dynamic in Mar Chiquita coastal lagoon (Argentina), and its relationship with potential nutrient sources. *Journal of Coastal Research, 21,* 818-825.

De Wolf, H. (1982). Method of coding of ecological data from diatoms for computer utilization. *Mededelingen Rijks Geologische Dienst, 36,* 95-99.

Debenay, J. P., Carbonel, P., Morzadec-Kerfourn, M.-T., Cazaboun, A., Denèfle, M. & Lézine, A. –M. (2003). Multi-bioindicator study of a small estuary in Vendée (France). *Estuarine, Coastal and Shelf Science, 58,* 843-860.

Denys, L. (1991/1992). *A check-list of the diatoms in the Holocene deposits of the Western Belgian coastal plain with the survey of their apparent ecological requirements, I. Introduction, ecological code and complete list.* Berchem: Service Geological of Belgium Professional Paper 246.

Denys, L. & De Wolf, H. (1999). Diatoms as indicators of coastal paleoenvironments and relative sea-level change. In E. F. Stoermer, & J. P. Smol (Eds.), *The diatoms: Applications for the Environmental and Earth Sciences* (pp. 277-297). London: Cambridge University Press.

Diodato, S.L. & Hoffmeyer, M.S. (2008). Contribution of planktonic and detritic fractions to the natural diet of mesozooplankton in Bahía Blanca estuary. *Hydrobiologia, 614,* 83-90.

Dodd, J.R. & Stanton, E.J. Jr. (1990). *Palaeoecology, Concepts and Applications,* Second Edition, Wiley-Interscience Publication, 553 pp.

EcoPlata Team (Eds.), 1996. *The Río de la Plata. An Environmental Overview.* An EcoPlata Project background report. Working draft, November 1996. Dalhouse University, Halifax, Nova Scotia, 242 pp.

Elgué, J.C., Bayseé, C., Burone, F. & Parietti, M. (1990). Distribución y sucesión especial del fitoplancton de superficie de la zona común de pesca Argentino-Uruguaya (Invierno de 1983). *Frente Marítimo, 6,* 67-107.

Escandel, A., Espinosa, M. A. & Isla, F. I. (2009). Diatomeas como indicadoras de variaciones de salinidad durante el Holoceno tardío en el estuario del río Negro, Patagonia Norte, Argentina. *Ameghiniana, 46,* 461-468.

Espinosa, M. A. (1988). Paleoecología de diatomeas del estuario del Río Quequén (Prov. de Buenos Aires, Argentina). *Thalassas, 6,* 33-44.

Espinosa, M. A. (1994). Diatom paleoecology of the Mar Chiquita lagoon delta, Argentina. *Journal of Paleolimnology, 10,* 17-23.

Espinosa, M. A. (1998). *Paleoecología de diatomeas en sedimentos cuaternarios del sudeste bonaerense,* Doctoral Thesis, Universidad Nacional de Mar del Plata.

Espinosa, M. A. (2001). Reconstrucción de paleoambientes holocenos de la costa de Miramar (provincia de Buenos Aires, Argentina) basada en diatomeas. *Ameghiniana, 38,* 27-34.

Espinosa, M. A., De Francesco, C. G. & Isla, F. I. (2003). Paleoenvironmental reconstruction of Holocene coastal deposits from the southeastern Buenos Aires province, Argentina. *Journal of Paleolimnology, 29,* 49-60.

Espinosa, M. A., Hassan, G. S. & Isla, F. I. (2006). Diatom distribution across a temperate microtidal marsh, Mar Chiquita coastal lagoon, Argentina. *Thalassas, 22,* 9-16.

Espinosa, M.A. (2008). Diatoms from Patagonia and Tierra del Fuego. In J. Rabassa (Ed.) *The Late Cenozoic of Patagonia and Tierra del Fuego* (pp. 383-392). Ciudad: Developments in Quaternary Sciences 11.

Fasano, J. L., Hernández, M. A., Isla, F. I. & Schnack, E. J. (1982). Aspectos evolutivos y ambientales de la laguna Mar Chiquita (provincia de Buenos Aires, Argentina). In P. Lasserre, & H. Postma (Eds.), *Coastal lagoons: Proceedings of the International Symposium of Coastal Lagoons* (pp. 285-292). Bordeaux: Oceanologica Acta 4.

Ferrando, H.J., de Castro, T.M. & Tenyn, E. (1964). Clave para las principales diatomeas planctónicas del Atlántico Sudoccidental, *Revista del Instituto de Investigaciones Pesqueras, 1,* 185-225.

Ferrario, M. E. (1972). Diatomeas pennadas de la ría de Puerto Deseado (provincia de Santa Cruz, Argentina). I. Araphidales. *Anales de la Sociedad Científica, 193,* 135-176.

Ferrario, M. E. (1981). Diatomeas centrales de la ría de Puerto Deseado (Santa Cruz, Argentina). IV.- S.O. Biddulphiineae, Fam. Eupodiscaceae y Fam. Lithodesmiaceae. *Darwiniana, 23,* 475-488.

Ferrario, M. E. (1984a). Diatomeas centrales de la ría de Puerto Deseado (Santa Cruz, Argentina). I.- S.O. Rhizosoleniineae Familia Rhizosoleniaceae y S.O. Biddulphiineae, Familia Chaetoceraceae. *Revista del Museo de La Plata, Nueva Serie 8, Botánica, 83,* 247-254.

Ferrario, M. E. (1984b). Diatomeas centrales de la ría de Puerto Deseado (Santa Cruz, Argentina). II.- S.O. Coscinodisciineae Familia Hemidiscaceae y Familia Melosiraceae. *Revista del Museo de La Plata, Nueva Serie 8, Botánica, 84,* 267-289.

Ferrario, M. E. (1984c). Diatomeas centrales de la ría de Puerto Deseado (Santa Cruz, Argentina). III.- S.O. Coscinodisciineae Familia Coscinodiscaceae, Familia Heliopeltaceae y Familia Thalassiosiraceae. *Revista del Museo de La Plata, Nueva Serie 8, Botánica, 85,* 291-311.

Ferrario, M.E. & Galván N.M. (1989). *Catálogo de las diatomeas marinas citadas entre los 36° y los 60° S con especial referencia al Mar Argentino*. Instituto Antártico Argentino, Publicación N°20, 327 pp. Buenos Aires.

Ferrario, M. E. & Sar, E. A. (1984). Diatomeas pennadas de la ría de Puerto Deseado (provincia de Santa Cruz). II. S.O. Raphidiineae. *Revista del Museo de La Plata, Nueva Serie 8, Botánica, 80,* 213-230.

Ferrario, M.E. & Sastre, V. (1990). Ultraestructura, polimorfismo y ecología de *Odontella aurita* (Lyngbye) Agardh (Bacillariophyceae) en el estuario del Río Chubut, Argentina. *Revista de la Facultad de Oceanografía, Pesca y Ciencias Alimentarias, 2,* 98-106.

Frenguelli, J. (1921). Los terrenos de la costa atlántica en los alrededores de Miramar (provincia de Buenos Aires) y sus correlaciones. *Boletín de la Academia de Ciencias de Córdoba, 24,* 325-485.

Frenguelli, J. (1923). Diatomeas de Tierra del Fuego. *Anales de la Sociedad Científica Argentina, 95,* 225-263.

Frenguelli, J. (1924). Diatomeas de Tierra del Fuego. *Anales de la Sociedad Científica Argentina, 98,* 5-89.

Frenguelli, J. (1925). Diatomeas de los arroyos del Durazno y Las Brusquitas en los alrededores de Miramar (provincia de Buenos Aires). *Physis, 8,* 129-185.

Frenguelli, J. (1935). Diatomeas de la Mar Chiquita al norte de Mar del Plata. *Revista del Museo de La Plata, Nueva serie 1, Botánica, 5,* 121-141.

Frenguelli, J. (1938). Diatomeas de la Bahía San Blas (provincia de Buenos Aires). *Revista del Museo de La Plata, Nueva serie 1, Botánica, 5,* 251-337.

Frenguelli, J. (1941). Diatomeas del Río de la Plata. *Revista del Museo de La Plata, Nueva serie 3, Paleontología, 16,* 77-251.

Frenguelli, J. (1945). Las diatomeas del Platense. *Revista del Museo de La Plata, Nueva serie 3, Botánica, 15,* 213-334.

Gayoso, A. M. (1981). *Estudio del fitoplancton del estuario de Bahía Blanca.* Bahía Blanca: Instituto Argentino de Oceanografía.

Gayoso, A. M. (1988). Variación estacional del fitoplancton de la zona más interna del estuario de Bahía Blanca (prov. Buenos Aires, Argentina). *Gayana, Botanica, 45,* 241-247.

Gayoso, A.M. (1996). Pytoplankton species composition and abundance off Río de la Plata (Uruguay). *Arch. Fish. Mar. Res., 44,* 257-265.

Gayoso, A. M. (1998). Long-term phytoplankton studies in the Bahía Blanca estuary, Argentina. *ICES Journal of Marine Science, 55,* 655-660.

Gayoso, A. M. (1999). Seasonal succession patterns of phytoplankton in the Bahía Blanca estuary (Argentina). *Botanica Marina, 42,* 367-375.

Gehrels, W. R., Roe, H. M. & Charman, D. J. (2001). Foraminifera, testate amoebae and diatoms as sea-level indicators in UK saltmarshes: a quantitative multiproxy approach. *Journal of Quaternary Science, 16,* 201-220.

Goldstein, S. T. & Watkins, G. T. (1999). Taphonomy of salt marsh foraminifera: an example of coastal Georgia. *Palaeogeography, Palaeoclimatology, Palaeoecology, 149,* 103-114.

Gómez, N. & Bauer, D. E. (1998a). Phytoplankton from the Southern Coastal Fringe of the Río de la Plata (Buenos Aires, Argentina). *Hydrobiologia, 380,* 1–8.

Gómez, N. & Bauer, D. E. (1998b). Coast phytoplankton of the Río de la Plata river and its relation to pollution. *International Association of Theoretical and Applied Limnology, 26,* 1032-1036.

Gómez, N. & Bauer, D. E. (2000). Diversidad fitoplanctónica en la franja costera sur del Río de la Plata. *Biología Acuática, 19,* 7-26.

Gómez, N., Hualde, P. R., Licursi, M. & Bauer, D. E. (2004). Spring phytoplankton of Río de la Plata: a temperate estuary of South America. *Estuarine, Coastal and Shelf Science, 61,* 301-309.

Gómez, N., Licursi, M. & Cochero, J. (2009). Seasonal and spatial distribution of the microbenthic communities of the Rio de la Plata estuary (Argentina) and possible environmental controls. *Marine Pollution Bulletin, 58,* 878-887.

Guarrera, S.A. (1950). Estudios hidrobiológicos en el Río de la Plata. *Instituto Nacional de Investigación de las Ciencias Naturales, Ciencias Botánicas 2,* 62 pp.

Guiry, M. D. & Guiry, G. M. (2009). AlgaeBase, World-wide electronic publication, National University of Ireland, Galway. http://www.algaebase.org; searched on July 2009.

Gutiérrez Tellez, B. M. & Schillizzi, R. A. (2002). Asociaciones de diatomeas en paleoambientes cuaternarios de la costa sur de la provincia de Buenos Aires, Argentina. *Pesquisas en Geociências, 29,* 59-70.

Hassan, G. S. (2008). *Diatomeas estuáricas del sudeste bonaerense: distribución, composición, diversidad y su aplicación en paleoecología.* Doctoral Thesis, Universidad Nacional de Mar del Plata.

Hassan, G. S., Espinosa, M. A. & Isla, F. I. (2006). Modern diatom assemblages in surface sediments from estuarine systems in the southeastern Buenos Aires Province, Argentina. *Journal of Paleolimnology, 35,* 39-53.

Hassan, G. S., Espinosa, M. A. & Isla, F. I. (2007). Dead diatom assemblages in surface sediments from a low impacted estuary: the Quequén Salado river, Argentina. *Hydrobiologia, 579,* 257-270.

Hassan, G. S., Espinosa, M. A. & Isla, F. I. (2008). Fidelity of dead diatom assemblages in estuarine sediments: how much environmental information is preserved? *Palaios, 23,* 112-120.

Hassan, G. S., Espinosa, M. A. & Isla, F. I. (2009). Diatom-based inference model for paleosalinity reconstructions in estuaries along the northeastern coast of Argentina. *Palaeogeography, Palaeoclimatology, Palaeoecology, 275,* 77-91.

Hentschel, E. (1932). Die biologischen methoden und das biologische beobachtungsmaterial der "meteor-expedition". *Wissenschaftl. Ergbn. d. deustch. atlant. Exped. a. d. Forsh. – Vermessungsschiff "Meteor" 1925-1927, 10,* 1-174.

Horton, B. P. (1999). The distribution of contemporary intertidal foraminifera at Cowpen Marsh, Tees Estuary, UK: implications for studies of Holocene sea-level changes. *Palaeogeography, Palaeoclimatology, Palaeoecology, 149,* 127-149.

Horton, B. P., Corbett, R., Culver, S. J., Edwards, R. J. & Hillier, C. (2006). Modern saltmarsh diatom distributions of the Outer Banks, North Carolina, and the development of a transfer function for high resolution reconstructions of sea level. *Estuarine, Coastal and Shelf Science, 69,* 381-394.

Hustedt, F. (1937/1938). *Systematische und ökologische Untersuchungen über die Diatomeen. "Flora von Java, Bali und Sumatra nach dem Material der Deutschen Limnologischen Sunda-Expedition"*, Archive für Hydrobiologie supplement, Band 15, 131-506.

Hustedt, F. (1953). Die Systematik der diatomeen in ihren beziehungen zur geologie und ökologie nebst einer revision des halobien-systems. *Svensk Botaniska Tidskr, 47,* 274-344.

Isla, F. I. & Espinosa, M. A. (2005). Holocene and historical evolution of an estuarine complex: the gravel spit of the Walker creek, Southern Buenos Aires. *XVI Congreso Geológico Argentino, Actas,* 149-154.

Isla, F. I., Fasano, J. L., Ferrero, L., Espinosa, M. A. & Schnack, E. J. (1986). Late Quaternary marine-estuarine sequences of the southeastern coast of Buenos Aires Province, Argentina. *Quaternary of South America and Antarctic Peninsula, 4,* 137-157.

Isla, F.I. & Bujalesky, G.G. (2004). Morphodynamics of a gravel-dominated macrotidal estuary: Rio Grande, Tierra del Fuego. *Revista de la Asociación Geológica Argentina, 59,* 220-228.

Juggins, S. (1992). *Diatoms in the Thames estuary, England: ecology, paleoecology, and salinity transfer function.* Berlin: J. Cramer.

Kidwell, S. M. (2001). Preservation of species abundance in marine death assemblages. *Science, 294,* 1091-1094.

Kidwell, S. M. (2002). Mesh-size effects on the ecological fidelity of death assemblages: a meta-analysis of molluscan live-dead studies. *Geobios, 24,* 107-119.

Kidwell, S. M. & Flessa, K. W. (1995). The quality of the fossil record: populations, species, and communities. *Annual Reviews of Ecology and Systematics, 26,* 269-299.

Kowalewski, M., Lazo, D. G., Carroll, M., Messina, C., Casazza, L., Puchalski, S., Gupta, N. S., Rothfus, T. A., Hannisdal, B., Sälgeback, J., Hendy, A., Stempien, J., Krause Jr., R., Terry, R. C., LaBarbera, M. & Tomašových, A. (2003). Quantitative fidelity of brachiopod-mollusk assemblages from modern subtidal environments of San Juan Islands, USA. *Journal of Taphonomy, 1,* 43-65.

Lange, K.B. (1985). Spatial and seasonal variations of diatom assemblages off the Argentinean coast (South Western Atlantic). *Oceanologica Acta, 8,* 361-369.

Licursi, M., Sierra, M. V. & Gómez, N. (2006). Diatom assemblages from a turbid coastal plain estuary: Río de la Plata (South America). *Journal of Marine Systems, 62,* 35–45.

Marini, M. F. & Piccolo, M. C. (2000). El balance hídrico en la cuenca del río Quequén Salado, Argentina. *Papeles de Geografía, 31,* 39-53.

Martínez Macchiavelo, J.C. (1979). Diatomeas del Océano Atlántico frente a la provincia de Buenos Aires, Argentina (Bacillariophyceae-Eupodiscales). *Revista del Museo Argentino de Ciencias Naturales Bernardino Rivadavia, Botánica 5,* 229-239.

Maynard, N. G. (1976). Relationship between diatoms in surface sediments of the Atlantic Ocean and the biological and physical oceanographic of the overlying waters. *Paleobiology, 2,* 99-121.

Metzeltin, D. & García-Rodríguez, F. (2003). *Las diatomeas Uruguayas.* DIRAC: Montevideo, Uruguay.

Moore, W. W. & McIntire, C. D. (1977). Spatial and seasonal distribution of littoral diatoms in Yanquina estuary, Oregon (USA). *Botanica Marina, 20,* 99-109.

Müller-Melchers, F.E. (1945). Diatomeas procedentes de algunas muestras de turba del Uruguay. *Comunicaciones Botánicas del Museo de Historia Natural de Montevideo, 1,* 1-25

Müller Melchers, F.E. (1952). Biddulphia chilensis Grev. as indicator of ocean currents. *Comunicaciones Botánicas del Museo de Historia Natural de Montevideo, 2,* 1-25.

Müller Melchers, F.E. (1953). New and little known diatoms from Uruguay and the South Atlantic coast. *Comunicaciones Botánicas del Museo de Historia Natural de Montevideo, 3,* 1-25.

Müeller Melchers, F. (1959). Plancton diatoms of the southern Atlantic Argentina and Uruguay coast. *Comunicaciones Botánicas del Museo de Historia Natural de Montevideo, 3,* 1-45.

Murray, J. W. & Pudsey, C. J. (2004). Living (stained) and dead foraminifera from the newly ice-free Larsen Ice Shelf, Weddell Sea, Antarctica: ecology and taphonomy. *Marine Micropaleontology, 53,* 67-81.

Ng, S. L. & Sin, F. S. (2003). A diatom model for inferring sea level change in the coastal waters of Hong Kong. *Journal of Paleolimnology, 30,* 427-440.

Oppenheim, D. R. (1991). Seasonal changes in epipelic diatoms along an intertidal shore, Berrow Flats, Somerset. *Journal of the Marine Biological Association of the United Kingdom, 71,* 579-596.

Parodi, E. R. & Barría de Cao, S. (2003). Benthic microalgal communities in the inner part of the Bahía Blanca estuary (Argentina): a preliminary qualitative study. *Oceanologica Acta, 25,* 279-284.

Paterson, D. M. (1989). Short-term changes in the erodibility of intertidal cohesive sediments related to the migratory behavior of epipelic diatoms. *Limnology and Oceanography, 34,* 223-324.

Perillo, G. M. E., Pérez, D. E., Piccolo, M. C., Palma, E. D. & Cuadrado, D. G. (2005). Geomorphologic and physical characteristics of a human impacted estuary: Quequén Grande River Estuary, Argentina. *Estuarine, Coastal and Shelf Science, 62,* 301-312.

Piccolo, M. C. & Perillo, G. M. (1999). The argentinean estuaries: a review. In G. M. E. Perillo, M. C. Piccolo, & M. Pino Quivira (Eds.), *Estuaries of South America: their geomorphology and dynamics* (pp. 101-132). Berlin: Springer-Verlag.

Piccolo, M. C., Perillo, G. M. & Arango, J. M. (1987). Hidrografía del estuario de Bahía Blanca, Argentina. *Geofísica, 26,* 75-89.

Popovich, C. A. (2004). Fitoplancton. In M. C. Piccolo, & M. Hoffmeyer (Eds.), *El ecosistema del estuario de Bahía Blanca* (pp. 69-78). Bahía Blanca: Instituto Argentino de Oceanografía.

Popovich, C. A. & Marcovecchio, J. E. (2008). Spatial and temporal variability of phytoplankton and environmental factors in a temperate estuary of South America (Atlantic coast, Argentina). *Continental Shelf Research, 28,* 236-244.

Popovich, C. A., Spetter, C. V., Marcovecchio, J. E. & Freije, R. H. (2008). Dissolved nutrient availability during winter diatom bloom in a turbid and shallow estuary (Bahía Blanca, Argentina). *Journal of Coastal Research, 24,* 95-102.

Pucci, A. E., Hoffmeyer, M. S., Freije, H. R., Barría, M. S., Popovich, C. A., Rusansky, C. & Asteasuain, R. (1996). Características de las aguas y del plancton en un sector del estuario del río Negro (República Argentina). In J.

Marcovecchio (Ed.), *Pollution Processes in Coastal Environments* (pp. 146-151). Mar del Plata: Universidad Nacional de Mar del Plata.

Resende, P., Azeiteiro, U. & Pereira, M. J. (2005). Diatom ecological preferences in a shallow temperate estuary (Ria de Aveiro, Western Portugal). *Hydrobiologia, 544,* 77-88.

Reta, R., Martos, P., Perillo, G. M. E., Piccolo, M. C. & Ferrante, A. (2001). Características hidrográficas del estuario de la laguna Mar Chiquita. In O.O. Iribarne (Ed.), *Reserva de Biosfera Mar Chiquita: características físicas, biológicas y ecológicas* (pp. 31-52). Mar del Plata: Editorial Martin.

Roggiero, M.F. (1988). Fitoplancton del Río de la Plata, I. *Lilloa, 37,* 137-152.

Rybarcyk, H. & Elkaïm, B. (2003). An analysis of the trophic network of a macrotidal estuary: the Seine Estuary (Eastern Channel, Normandy, France). *Estuarine, Coastal and Shelf Science, 58,* 775-791.

Rzeznik-Orignac, J., Fichet, D. & Boucher, G. (2003). Spatio-temporal structure of the nematode assemblages of the Brouage mudflat (Marennes Oléron, France). *Estuarine, Coastal and Shelf Science, 58,* 77-88.

Sar, E. A., Sala, S. E., Sunesen, I., Henninger, M.S. & Montastruc, M. (2009). Catalogue of the genera, species, and infraspecific taxa erected by J. Frenguelli. *Diatom Monographs, 10,* 419 pp.

Santinelli, N., Sastre, A. V. & Caille, G. (1990). Fitoplancton del estuario inferior del río Chubut y su relación con la salinidad y la temperatura. *Revista de la Asociación de Ciencias Naturales del Litoral, 21,* 69-79.

Sastre, A. V., Santinelli, N. H. & Caille, G. (1990). Diatomeas y dinoflagelados del estuario del río Chubut (Patagonia, Argentina). II. Estructura de las comunidades. *Revista Fac. de Ocean., Pesq. y Cs Alimentarias, 2,* 181-192.

Sastre, A. V., Santinelli, N. H. & Sendin, M. (1994). Floración de *Aulacoseira granulata* (Ehr.) Simonsen (Bacillariophyceae) en el curso inferior del río Chubut. *Revista Brasileira de Biologia, 54,* 641-647.

Sastre, A.V., Santinelli, N.H., Otaño, S.H. & Ivanissevich, M.E. (1998). Water quality in the lower section of the Chubut River, Patagonia, Argentina. *Verhandlungen Internationale Vereiningen Limnolgie, 26,* 951-955.

Sawai, Y. (2001). Distribution of living and dead diatoms in tidal wetlands of northern Japan: relations to taphonomy. *Palaeogeography, Palaeoclimatology, Palaeoecology, 173,* 125-141.

Sawai, Y., Horton, B. P. & Nagumo, T. (2004). The development of a diatom-based transfer function along the Pacific coast of eastern Hokkaido, northern Japan – an aid in paleoseismic studies of the Kuril subduction zone. *Quaternary Science Reviews, 23,* 2467-2483.

Schillizzi, R., Gutiérrez Tellez, B. & Aramayo, S. (2006). Reconstrucción paleoambiental del Cuaternario en las barrancas del río Quequén Salado, provincia de Buenos Aires, Argentina. *III Congreso Argentino de Cuaternario y Geomorfología, Actas,* 649-658.

Schmidt, A., Schmidt, M., Fricke, F., Heiden, H., Müller, O. & Hustedt, F. 1874-1959. *Atlas der Diatomaceen Kuncle.* Serie I-X, Leipzig.

Schwindt, E., Iribarne, O.O. & Isla, F. I. (2004). Physical effects of an invading reef-building polychaete on an Argentinean estuarine environment. *Estuarine, Coastal and Shelf Science, 59,* 109-120.

Scott, D. B. & Medioli, F. S. (1980). Living vs. total foraminiferal populations: their relative usefulness in paleoecology. *Journal of Paleontology, 54,* 814-831.

Sherrod, B. L. (1999). Gradient analysis of diatom assemblages in a Puget Sound salt marsh: can such assemblages be used for quantitative paleoecological reconstructions? *Palaeogeography, Palaeoclimatology,* Palaeoecology, 149, 213-226.

Simonsen, R. (1969). Diatoms as indicators in estuarine environments. *Veröffentl. Inst. Meersforsch. Bremerhaven, 11,* 287-291.

SMEBD (2009). WoRMS: The World Register of Marine Species. Available online at http://www.marinespecies.org. Accessed on July 2009.

Snoeijs, P. (1999). Diatoms and environmental change in brackish waters. In E. F. Stoermer, & J. P. Smol (Eds.). *The diatoms: Applications for the environmental and Earth Sciences* (pp. 298-333). London: Cambridge University Press.

Sylvestre, F., Beck-Eichier, B., Duleba, W. & Debenay, J. (2001). Modern benthic diatom distribution in a hypersaline coastal lagoon: the Lagoa de Arauama (R.J.), Brazil. *Hydrobiologia, 443,* 213-231.

Tempère, M. & Peragallo, H. (1907-1915). *Diatomees du Monde Entier.* Edition 2, 30 fasc., Arcachon, Graz-sur-Loing.

ter Braak, C. J. F. & Juggins, S. (1993). Weighted averaging partial least squares regression (WA-PLS): an improved method for reconstructing environmental variables from species assemblages. *Hydrobiologia, 269/270,* 485-502.

ter Braak, C. J. F., Juggins, S., Birks, H. J. B. & van der Voet, H. (1993). Weighted averaging partial least squares regression (WA-PLS): definition and comparison with other methods for species-environment calibration. In: G. P. Patil, & C. R. Rao (Eds.), *Multivariate Environmental Statistics* (pp. 525-560). Amsterdam: Elsevier Science Publishers.

Thiemann, K. (1934). Das Plankton der Flussmündungen. *Biologische Sonderuntersuchungen Wissensch. Ergebn. Atlant. Exped. Forsch. Vermessungsschiff "Meteor", 12,* 199-273.
Underwood, G. J. C. (1994). Seasonal and spatial variation in epipelic diatom assemblages in the Severn Estuary. *Diatom Research, 9,* 451-472.
Underwood, G. J. C. (1997). Microalgal colonization in a saltmarsh restoration scheme. *Estuarine, Coastal and Shelf Science, 44,* 471-481.
Underwood, G. J. C. & Paterson, D. M. (1993). Seasonal changes in diatom biomass, sediment stability and biogenic stabilization in the Severn Estuary. *Journal of the Marine Biological Association of the United Kingdom, 73,* 871-887.
Van Dam, H., Mertens, A. & Sinkeldam, J. (1994). A coded checklist and ecological indicator values of freshwater diatoms from The Netherlands. *Netherlands Journal of Aquatic Ecology, 28,* 117-133.
Villafañe, V., Hebling, E. W. & Santamarina, J. (1991). Phytoplankton blooms in the Chubut river estuary (Argentina): influence of stratification and salinity. *Revista de Biología Marina (Valparaíso), 26,* 1-20.
Vos, P. C. & De Wolf, H. (1988). Methodological aspects of paleo-ecological diatom research in coastal areas of the Netherlands. *Geologie en Mijnbouw, 67,* 31-40.
Vos, P. C. & De Wolf, H. (1993). Diatoms as a tool for reconstructing sedimentary environments in coastal wetlands; metodological aspects. *Hydrobiologia, 269,* 285-296.
Vouilloud, A. A. (2003). *Catálogo de diatomeas continentales y marinas de Argentina.* La Plata: Asociación Argentina de Ficología.
Witkowski, A., Cedro, B., Kierzek, A. & Baranowski, D. (2009). Diatoms as a proxy in reconstructing the Holocene environmental changes in the southwestern Baltic Sea: the lower Rega River Valley sedimentary record. *Hydrobiologia, 631,* 155-172.
Wolfstein, K., Colijn, F. & Doerffer, R. (2000). Seasonal dynamics of microphytobenthos biomass and photosynthetic characteristics in the Northern German Wadden Sea, obtained by the photosynthetic light dispensation system. *Estuarine, Coastal and Shelf Science, 51,* 651-662.
Ysebaert, T., Herman, P. M. J., Meire, P., Craeymeersch, J., Verbeek, H. & Heip, C. H. R. (2003). Large-scale spatial patterns in estuaries: estuarine macrobenthic communities in the Schelde estuary, NW Europe. *Estuarine, Coastal and Shelf Science, 57,* 335-355.

Zárate, M. A., Espinosa, M. A. & Ferrero, L. (1998). Palaeoenvironmental implications of a Holocene diatomite, Pampa Interserrana, Argentina. *Quaternary of South America and Antarctic Peninsula, 5,* 135-152.

Zong, Y. & Horton, B. P. (1999). Diatom-based tidal-level transfer functions as an aid in reconstructing Quaternary history of sea-level movements in the UK. *Journal of Quaternary Science, 14,* 153-167.

Reviewed by:

**Dr. Marcela A. Espinosa,** CONICET-Universidad de Mar del Plata, Instituto de Geología de Costas y del Cuaternario, CC 722 7600 Mar del Plata, ARGENTINA, e-mail: maespin@mdp.edu.ar.

**Dr. Felipe García-Rodríguez,** Universidad de la República, Facultad de Ciencias, Montevideo, URUGUAY, e-mail: Felipe.garciarodriguez@gmail.com.

# APPENDIX I.

List of the recent publications containing information about estuarine diatoms from Argentina used to construct Appendix II, type of sample analyzed (P: plankton; S: sediment; E: epiphytes under vegetation), and number of taxa mentioned (n/a: not available). Works providing of environmental data are marked (+).

| Estuary | Author | Sample | N° of taxa | Env. Data |
|---|---|---|---|---|
| Rio de la Plata | Gómez and Bauer (1998) | P | 32 | + |
|  | Carreto et al. (2003) | P | 8 | + |
|  | Metzeltin & García-Rodríguez (2003) | S | 295 | - |
|  | Gómez et al. (2004) | P | 15 | + |
|  | Calliari et al. (2005) | P | 10 | + |
|  | Licursi et al. (2006) | P | 87 | + |
|  | Bauer et al. (2007) | E | 44 | + |
|  | Carreto et al. (2008) | P | 4 | + |
|  | Calliari et al. (2009) | P | 11 | + |
|  | Gómez et al. (2009) | S | 52 | + |
| Mar Chiquita coastal lagoon | De Marco (2002) | P | n/a | + |
|  | De Marco et al. (2005) | P | n/a | + |
|  | Espinosa et al. (2006) | S | 20 | + |
|  | Hassan et al. (2006) | S | 31 | + |
|  | Hassan et al. (2008) | S | 15 | + |
|  | Hassan et al. (2009) | S | 28 | + |
| Rio Quequén Grande | Hassan et al. (2006) | S | 37 | + |
|  | Hassan et al. (2008) | S | 18 | + |
|  | Hassan et al. (2009) | S | 36 | + |

| | | | | |
|---|---|---|---|---|
| **Rio Quequén Salado** | Hassan et al. (2007) | S | 32 | + |
| | Hassan et al. (2009) | S | 30 | + |
| **Bahía Blanca** | Gayoso (1981) | P | 30 | + |
| | Gayoso (1988) | P | 23 | + |
| | Gayoso (1998) | P | n/a | + |
| | Gayoso (1999) | P | 19 | + |
| | Andrade et al. (2000) | P | 1 | + |
| | Parodi and Barría de Cao (2003) | S | 13 | + |
| | Parodi (2004) | S | 12 | + |
| | Popovich (2004) | P | 48 | + |
| | Diodato and Hoffmeyer (2008) | P | 13 | + |
| | Popovich et al. (2008) | P | 20 | + |
| | Popovich and Marcovechio (2009) | P | 14 | + |
| **Rio Negro** | Pucci et al. (1996) | P | 19 | + |
| **Bahía San Blas** | Isla and Espinosa (2005) | S | 15 | + |
| **Río Chubut** | Ferrario and Sastre (1990) | P | 1 | + |
| | Sastre et al. (1990) | P | 40 | + |
| | Santinelli et al. (1990) | P | 39 | + |
| | Villafañe et al. (1991) | P | 12 | + |
| | Sastre et al. (1994) | P | 1 | + |
| | Ayestarán and Sastre (1995) | P | 28 | - |
| | Sastre et al. (1998) | P | 10 | + |
| **Ría Puerto Deseado** | Ferrario (1972, 1981, 1984 a, b) Ferrario and Sar (1984) | P | 88 | + |

# APPENDIX II

List of the diatom taxa cited for estuaries of Argentina, based on the publications listed in Appendix I. 1: Río de la Plata; 2: Mar Chiquita coastal lagoon; 3: Río Quequén Grande; 4: Río Quequén Salado; 5: Bahía Blanca; 6: Río Negro; 7: Bahía San Blas; 8: Río Chubut; 9: Ría Puerto Deseado. Letters indicate the type of sample in which each taxon was found; P: plankton; S: sediment; E: epiphytes under vegetation. In the last column a summary of the salinity ranges reported in all works is provided. WA: optima ± tolerance calculated by weighted averaging; n/a: no data available.

| TAXA NAME/AUTHORITY | 1 | 2 | 3 | 4 | 5 | 6 | 7 | 8 | 9 | SALINITY |
|---|---|---|---|---|---|---|---|---|---|---|
| *Achnanthes brevipes* Agardh | | S | S | S | | | S | | | 9-38‰ (marine/brackish, euryhaline) |
| *Achnanthes brevipes* var. *intermedia* (Kützing) Cleve | | | | | | | | P | | 32-34‰ (marine) |
| *Achnanthes elata* (Leud.-Fort.) Gandhi | S | | | | | | | | | n/a |
| *Achnanthes exigua* Grunow | S | | | | | | | | | n/a |
| *Achnanthes inflata* (Kützing) Grunow | P/S | | | | | | | | | 0-0.8‰ (freshwater) |
| *Achnanthes inflata* var. *gibba* Gandhi | S | | | | | | | | | n/a |
| *Achnanthes inflatagrandis* Metzeltin, Lange-Bertalot & Garcia-Rodriguez | S | | | | | | | | | n/a |
| *Achnanthes lacus-vulcani* Lange-Bertalot & Krammer | | | | | | | S | | | 38‰ |
| *Achnanthes reversa* Lange-Bertalot | | S | S | | | | | | | WA: 19.5±5.5 (marine/brackish) |
| *Achnanthes subelata* Metzeltin, Lange-Bertalot & Garcia-Rodriguez | S | | | | | | | | | n/a |
| *Achnanthidium biasolettianum* (Grunow) Round & Bukht. | | S | | | | | | | | n/a |
| *Achnanthidium coarctatum* Brébisson ex Smith | S | | | | | | | | | n/a |
| *Achnanthidium lanceolatum* spp. *biporoma* | S | | | | | | | | | n/a |

(Continued)

| TAXA NAME/AUTHORITY | 1 | 2 | 3 | 4 | 5 | 6 | 7 | 8 | 9 | SALINITY |
|---|---|---|---|---|---|---|---|---|---|---|
| *Achnanthidium lanceolatum* spp. *miota* Lange-Bertalot | S | | | | | | | | | n/a |
| *Achnanthidium lanceolatum* spp. *frequentissima* Lange-Bertalot | S | | | | | | | | | n/a |
| *Achnanthidium minutissimum* (Kützing) Czarnecki | S | S | S | S | | | | | | WA: 4.8±7.7 (brackish, euryhaline) |
| *Achnanthidium parexigua* (Metzeltin & Lange-Bertalot) Metzeltin | S | | | | | | | | | n/a |
| *Actinocyclus* spp. | | | | | | P | | | | - |
| *Actinocyclus actinochilus* (Ehrenberg) Simonsen | | | | | | | | P | | n/a |
| *Actinocyclus divisus* (Grunow) Hustedt | | S | S | | | | | | | 8.4‰ (brackish) |
| *Actinocyclus kutzingii* (A. Schmidt) Simonsen | | | | | P | | | | | n/a |
| *Actinocyclus normanii* (Gregory) Hustedt | P/S/E | | | | | | | | | 0-18‰ (brackish/marine) |
| *Actinocyclus normanii* f. *subsalsus* (Juhlin-Dannfelt) Hust. | S | | | | | | | | | n/a |
| *Actinocyclus octonarius* Ehrenberg | P | S | | | | | | | P | 0-34‰ (marine/brackish, euryhaline) |
| *Actinocyclus subocellatus* (Grunow) Rattray | | | | | | | | | P | 32-34‰ (marine) |
| *Actinocyclus subtilis* (Gregory) Ralfs | | | | | | | | | P | 32-34‰ (marine) |
| *Actinoptychus* spp. | P | | | | | P | | P | | - |
| *Actinoptychus adriaticus* Grunow | | | | | P | | | | | n/a |
| *Actinoptychus campanulifer* Schmidt | | | | | | | | | P | 32-34‰ (marine) |
| *Actinoptychus frenguellii* Müller Melchers | | | | | | | | | P | 32-34‰ (marine) |
| *Actinoptychus senarius* (Ehrenberg) Ehrenberg | P | | | | P | | | P | P | 0-34‰ (marine/brackish, euryhaline) |
| *Actinoptychus splendens* (Shadbolt) Ralfs | | S | | S | P | | | | P | 32-34‰ (marine) |

(Continued)

| TAXA NAME/AUTHORITY | 1 | 2 | 3 | 4 | 5 | 6 | 7 | 8 | 9 | SALINITY |
|---|---|---|---|---|---|---|---|---|---|---|
| *Actinoptychus splendens* var. *glabrata* (Grunow) Pantocsek | | | | | | | | | P | 32-34‰ (marine) |
| *Actinoptychus vulgaris* Schumann | | | | | | | | P | P | 25-35‰ (marine/brackish) |
| *Amphipleura lindheimeri* Grunow | S | | | | | | | | | n/a |
| *Amphipleura pellucida* (Kützing) Kützing | S | | | | | | | | | n/a |
| *Amphipleura rutilans* var. *antarctica* (Grunow) Grunow | | | | | | | | | P | 32-34‰ (marine) |
| *Amphitetras antediluviana* Ehrenberg | S | | | | | | | | | n/a |
| *Amphora* spp. | | S | S | S | | | | | | - |
| *Amphora acutiuscula* Kützing | S | S | S | S | | | | | | WA: 28±3‰ (marine/brackish) |
| *Amphora coffeaeformis* (Agardh) Kützing | | S | S | S | | | | | | WA: 5.5±5.75‰ (brackish/freshwater) |
| *Amphora commutata* Grunow | S | | | | | | | | | n/a |
| *Amphora exigua* Gregory | | | | | | | | | P | 32-34‰ (marine) |
| *Amphora frenguelli* Forti | | S | S | S | | | | | | 0.5-3‰ (brackish/freshwater) |
| *Amphora helenensis* Giffen | | S | S | S | | | | | | WA: 9.5±6.5‰ (brackish/marine) |
| *Amphora libyca* Ehrenberg | S | S | S | S | | | | | | WA: 4.5±5.5‰ (brackish/freshwater) |
| *Amphora montana* Krasske | S | S | S | | | | | | | <1‰ (freshwater) |
| *Amphora normanii* Rabenhorst | S | | | | | | | | | n/a |
| *Amphora ovalis* (Kützing) Kützing | | | S | S | | | | | | 30-40‰ (marine) |
| *Amphora pediculus* (Kützing) Grunow | | S | S | | | | | | | WA: 2.8±4.6‰ (brackish) |
| *Amphora veneta* Kützing | | S | S | S | | | | | | WA: 2.6±3‰ (brackish) |
| *Anomoeoneis sphaerophora* Pfitzer | | S | S | S | | | | | | 0-2‰ (brackish/freshwater) |
| *Asterionella formosa* Hassall | | | | | | P | | P | | 0-26‰ (marine/brackish, |

(Continued)

| TAXA NAME/AUTHORITY | 1 | 2 | 3 | 4 | 5 | 6 | 7 | 8 | 9 | SALINITY |
|---|---|---|---|---|---|---|---|---|---|---|
| *Asterionellopsis glacialis* (Castracane) Round | | | | | P | P | | | P | 26-34‰ (marine) |
| *Aulacoseira ambigua* (Grunow) Simonsen | P | | | | | | | | | 0-15‰ (brackish/freshwater) |
| *Aulacoseira distans* (Ehrenberg) Simonsen | P/S | S | | | | | | | | 0-15‰ (brackish/freshwater) |
| *Aulacoseira granulata* (Ehrenberg) Simonsen | P/S/E | S | S | | P | P | | P | | 0-15‰ (brackish/freshwater) |
| *Aulacoseira granulata* var. *angustissima* (Müller) Simonsen | P | | | | | | | | | 0-15‰ (brackish/freshwater) |
| *Aulacoseira granulata* var. *valida* (Hustedt) Simonsen | S | | | | | | | | | n/a |
| *Aulacoseira italica* (Ehrenberg) Simonsen | | | | | | | | | P | 32-34‰ (marine) |
| *Aulacoseira muzzanensis* (Meister) Krammer | P | | | | | | | | | 0-0.2‰ (freshwater) |
| *Auliscus sculptus* (Smith) Ralfs | | | S | | | | | | P | 32-34‰ (marine) |
| *Bacillaria paradoxa* Gmelin | | S | S | S | | | | | | 0.5-14‰ (brackish/freshwater) |
| *Bacteriastrum furcatum* Shadbolt | | | | | | | | P | | n/a |
| *Berkeleya rutilans* (Trentepohl) Grunow | | S | S | S | | | | | P | 32-34‰ (marine) |
| *Biddulphia alternans* (Bailey) Van Heurck | | | | | | | | P | P | 0-35‰ (marine/brackish, euryhaline) |
| *Biddulphia antediluviana* (Ehrenberg) Van Heurck | | | | | | | | P | P | 0-35‰ (marine/brackish, euryhaline) |
| *Biddulphia biddulphiana* (Smith) Boyer | S | | | | | | | | | n/a |
| *Biddulphia rhombus* (Ehrenberg) Smith | | S | | | | | | | | 24‰ (marine/brackish) |
| *Brachysira neoexilis* Lange-Bertalot | S | | | | | | | | | n/a |
| *Brebissonia lanceolata* (Agardh) Mahoney & Reimer | | S | | S | | | | | | 0-5‰ (brackish/freshwater) |
| *Caloneis amphisbaena* (Bory) Cleve | | | | | | | | P | | <1‰ (freshwater) |

(Continued)

| TAXA NAME/AUTHORITY | 1 | 2 | 3 | 4 | 5 | 6 | 7 | 8 | 9 | SALINITY |
|---|---|---|---|---|---|---|---|---|---|---|
| *Caloneis brevis* | S | | | | | | | | | n/a |
| *Caloneis hyalina* Hustedt | S | | | | | | | | | n/a |
| *Caloneis permagna* (Bailey) Cleve | | S | | S | | | | | | 0-3‰ (brackish/freshwater) |
| *Caloneis tenuis* (Gregory) Krammer | S | | | | | | | | | n/a |
| *Caloneis westii* (Smith) Hendey | | S | S | S | | | | P | | <1‰ (freshwater) |
| *Campilosira* spp. | | S | | S | | | | | | - |
| *Campylodiscus clypeus* (Ehrenberg) Ehrenberg | | | S | S | | | | | | 0-7‰ (brackish/freshwater) |
| *Capartogramma crucicula* (Grunow) Ross | S | | | | | | | | | n/a |
| *Catacombus gaillonii* (Bory) Williams & Round | S | | | | | | | | P | 32-34‰ (marine) |
| *Catenula adhaerens* (Mereschkowsky) Mereschkowsky | | S | S | S | | | | | | WA: 20±5.4‰ (marine/brackish) |
| *Cavinula lapidosa* (Krasske) Lange-Bertalot | S | | | | | | | | | n/a |
| *Cavinula monoculata* (Hustedt) Mann | S | | | | | | | | | n/a |
| *Cerataulina pelagica* (Cleve) Hendey | | | | | P | | | | | 23-30‰ (marine/brackish) |
| *Chaetoceros* spp. | P | | | | P | | | P | | - |
| *Chaetoceros affinis* Lauder | P | | | | | | | | | 5-18‰ (brackish/marine) |
| *Chaetoceros brevis* Schütt | P | | | | | | | | | 17.5-18.5 (brackish/marine) |
| *Chaetoceros ceratosporus* Ostenfeld | | | | | P | | | | | 30.4-32.8‰ (marine) |
| *Chaetoceros ceratosporus* var. *brachysetus* Rines & Hargr. | | | | | P | | | | | 30.4-32.8‰ (marine) |
| *Chaetoceros convolutus* Castracane | | | | | | | | | P | 32-34‰ (marine) |
| *Chaetoceros debilis* Cleve | | | | | P | | | | | 20-33‰ (marine/brackish) |
| *Chaetoceros decipiens* Cleve | | | | | | | | | P | 32-34‰ (marine) |

(Continued)

| TAXA NAME/AUTHORITY | 1 | 2 | 3 | 4 | 5 | 6 | 7 | 8 | 9 | SALINITY |
|---|---|---|---|---|---|---|---|---|---|---|
| *Chaetoceros diadema* (Ehrenberg) Gran | | | | | P | | | | | 30.4-32.8‰ (marine) |
| *Chaetoceros similis* Cleve | | | | | P | | | | P | 25-34‰ (marine/brackish) |
| *Chaetoceros socialis* Lauder | | | | | | | | | P | 32-34‰ (marine) |
| *Chaetoceros subtilis* Cleve | P | | | | P | | | | P | 18-40‰ (marine/brackish) |
| *Chaetoceros subtilis* var. *abnormis* Prosckina-Lavrenko | | | | | P | | | | | n/a |
| *Chaetoceros teres* Cleve | | | | | | | | | P | 32-34‰ (marine) |
| *Cocconeis* spp. | | | | | S/P | | | P | | - |
| *Cocconeis grunowii* Pantocsek | | | | | | | | | P | 32-34‰ (marine) |
| *Cocconeis guttata* Hustedt & Aleem | | | | | | | S | | | 38‰ (marine) |
| *Cocconeis neodiminuta* Krammer | S | S | | | | | | | P | 32-38‰ (marine) |
| *Cocconeis pediculus* Ehrenberg | S | | | | | | | | | n/a |
| *Cocconeis pellucida* var. *minor* Grunow | | | | | | | | | P | 32-34‰ (marine) |
| *Cocconeis placentula* Ehrenberg (+ vars.) | P | S | S | S | | | | P | P | 0-34‰ (freshwater to marine, euryhaline) |
| *Cocconeis scutellum* Ehrenberg | | | | | | | | | P | 32-34‰ (marine) |
| *Cocconeis scutellum* var. *parva* (Grunow) Cleve | | | | | | | S | | | 38‰ (marine) |
| *Corethron criophilum* Castracane | | | | | P | | | | | n/a |
| *Coscinodiscus* spp. | P | | | | | P | | | | - |
| *Coscinodiscus argus* Ehrenberg | S | | | | | | | | | n/a |
| *Coscinodiscus asteromphalus* Ehrenberg | S | | | | | | | | | n/a |
| *Coscinodiscus bispculptus* Rattray | S | | | | | | | | | n/a |
| *Coscinodiscus concinnus* Smith | | | | | | | | | P | 32-34‰ (marine) |

*(Continued)*

| TAXA NAME/AUTHORITY | 1 | 2 | 3 | 4 | 5 | 6 | 7 | 8 | 9 | SALINITY |
|---|---|---|---|---|---|---|---|---|---|---|
| *Coscinodiscus curvatulus* Grunow | | S | | | | | | P | P | 32-34‰ (marine) |
| *Coscinodiscus granii* Gough | | | | | P | | | | | n/a |
| *Coscinodiscus janischii* Schmidt | | | | | | | | | P | 32-34‰ (marine) |
| *Coscinodiscus jonesianus* (Greville) Ostenfeld | | | | | | | | | P | 32-34‰ (marine) |
| *Coscinodiscus marginato-lineatus* var. *antarctica* Manguin | | | | | | | | | P | 32-34‰ (marine) |
| *Coscinodiscus marginatus* Ehrenberg | | | | | P | | | | P | 32-34‰ (marine) |
| *Coscinodiscus nitidus* Gregory | | | | | | | | | P | 32-34‰ (marine) |
| *Coscinodiscus obscurus* Schmidt | | | | | | | | | P | 32-34‰ (marine) |
| *Coscinodiscus oculus-iridis* (Ehrenberg) Ehrenberg | | | | | P | | | | P | 32-34‰ (marine) |
| *Coscinodiscus perforatus* var. *cellulosa* Grunow | | | | | | | | | P | 32-34‰ (marine) |
| *Coscinodiscus radiatus* Ehrenberg | P | S | S | | P | | | | P | 32-34‰ (marine) |
| *Coscinodiscus rothii* (Ehrenberg) Grunow | | | | | | | | | P | 32-34‰ (marine) |
| *Coscinodiscus rothii* var. *subsalsum* (Juhlin-Dann.) Hustedt | | S | | | | | | | | n/a |
| *Cosmioneis pusilla* var. *incognita* (Krasske) Aboal | S | | | | | | | | | n/a |
| *Craticula accomoda* (Hustedt) Mann | P | | | | | | | | | 0.1-0.4‰ (freshwater) |
| *Craticula ambigua* (Ehrenberg) Mann | S | | | | | | | | | n/a |
| *Craticula cuspidata* (Kutzing) Mann | P/S/E | S | | S | | | | P | | <1‰ (freshwater) |
| *Craticula halophila* (Grunow) Mann | P/S/E | | | | | | | | | <1‰ (freshwater) |
| *Craticula pampeana* (Frenguelli) Lange-Bertalot | S | | | | | | | | | n/a |
| *Craticula submolesta* (Hustedt) Lange-Bertalot | S | | | | | | | | | n/a |
| *Ctenophora pulchella* (Ralfs) Williams & Round | S | | | | | | | | | n/a |

(Continued)

| TAXA NAME/AUTHORITY | 1 | 2 | 3 | 4 | 5 | 6 | 7 | 8 | 9 | SALINITY |
|---|---|---|---|---|---|---|---|---|---|---|
| *Cyclotella* spp. | P | | | | P | P | | | | - |
| *Cyclotella atomus* Hustedt | P | | | | P | | | | | 0-0.2‰ (freshwater) |
| *Cyclotella meneghiniana* Kützing | P/S/E | S | S | S | P | | | P | P | WA: 6.3±7.5‰ (brackish, euryhaline) |
| *Cyclotella striata* (Kützing) Grunow | P | S | S | S | | | | | | 0-15‰ (brackish/freshwater) |
| *Cyclotella stylorum* Brightwell | | S | | | | | S | | | 38‰ (marine) |
| *Cylindrotheca closterium* (Ehrenberg) Reimann & Lewin | | | | | S/P | | | | P | 30-36‰ (marine) |
| *Cymatopleura solea* (Brébisson) W. Smith | | | | | | | | P | | 0-10‰ (freshwater/brackish) |
| *Cymatosira belgica* Grunow | | S | S | S | | | S | | | WA: 19.8±8‰ (marine/brackish) |
| *Cymbella* spp. | | | | | | | | P | | - |
| *Cymbella affinis* Kützing | P | | S | | | | | P | | 0-0.2‰ (freshwater) |
| *Cymbella australica* (Schmidt) Cleve | S | | | | | | | | | n/a |
| *Cymbella cistula* (Hemprich & Ehrenberg) Kirchner | | S | | S | | | | P | | 0-10 (freshwater/brackish) |
| *Cymbella cymbiformis* Agardh | | S | S | S | | | | | | 0-6‰ (freshwater/brackish) |
| *Cymbella cymbiformis* var. *nonpunctata* Fontell | | | | | | | | P | | <1‰ (freshwater) |
| *Cymbella neocistula* Krammer | S | | | | | | | | | n/a |
| *Cymbella prostrata* (Berkeley) Cleve | | | | | | | | P | | <1‰ (freshwater) |
| *Cymbella proxima* Patrick & Reimer | S | | | | | | | | | n/a |
| *Cymbella tumida* (Brébisson) Van Heurk | | | | | | | | P | | <1‰ (freshwater) |
| *Cymbella turgidula* Grunow | S | | | | | | | | | n/a |
| *Cymbopleura naviculiformis* (Auerswald) Krammer | S | | | | | | | | | n/a |
| *Dactyliosolen fragilissimus* (Bergon) Hasle | | | | | | | | | P | 32-34‰ (marine) |

| TAXA NAME/AUTHORITY | 1 | 2 | 3 | 4 | 5 | 6 | 7 | 8 | 9 | SALINITY |
|---|---|---|---|---|---|---|---|---|---|---|
| *Delicata nepouiana* Krammer | S | | | | | | | | | n/a |
| *Denticula elegans* Kützing | | S | S | S | | | | | | 0-3‰ (freshwater/brackish) |
| *Denticula kuetzingii* Grunow | | S | S | S | | | | | | WA: 3.8±5‰ (freshwater/brackish) |
| *Denticula tenuis* Kützing | | | S | S | | | | | | 0-2‰ (freshwater/brackish) |
| *Denticula valida* (Pedicino) Grunow | S | | | | | | | | | n/a |
| *Diadesmis contenta* (Grunow) Mann | S | | | | | | | | | n/a |
| *Diatoma moniliformis* Kützing | | | S | | | | | | | <1‰ (freshwater) |
| *Diatoma vulgaris* Bory | | S | S | S | | | | | | 0-5‰ (freshwater/brackish) |
| *Dickieia subinflata* (Grunow) Mann | | S | S | S | | | | | | 15-21‰ (marine/brackish) |
| *Dimeregramma minor* (Gregory) Ralfs | | S | S | S | | | | | | WA: 20±5‰ (marine/brackish) |
| *Diploneis caffra* (Giffen) Witkowski | S | | | | | | | | | n/a |
| *Diploneis chilensis* (Hustedt) Lange-Bertalot | S | | | | | | | | | n/a |
| *Diploneis interrupta* (Kützing) Cleve | | S | S | S | | | | | | 8.3-29‰ (marine/brackish) |
| *Diploneis ovalis* (Hilse) Cleve | | S | S | S | | | | | | 0.5-4‰ (freshwater/brackish) |
| *Diploneis puella* (Schumann) Cleve | | S | S | S | | | | | | 0.5-28‰ (marine/brackish) |
| *Diploneis smithii* (Brébisson) Cleve | | S | S | | | | | | | 0.5-2‰ (freshwater/brackish) |
| *Diploneis smithii* var. *constricta* Heiden | | S | | | | | | | | n/a |
| *Diploneis subovalis* Cleve | S | | | | | | | | | n/a |
| *Discostella pseudostelligera* (Hustedt) Houk & Klee | P | | | | | | | | | 0-0.2‰ (freshwater) |
| *Ditylum brightwellii* (West) Grunow | | | | | P | | | P | P | 30.4-34‰ (marine) |
| *Ditylum sol* (Schmidt) Cleve | | | | | | | | | P | 32-34‰ (marine) |

(Continued)

| TAXA NAME/AUTHORITY | 1 | 2 | 3 | 4 | 5 | 6 | 7 | 8 | 9 | SALINITY |
|---|---|---|---|---|---|---|---|---|---|---|
| *Encyonema mesiana* (Cholnoky) Krammer | S | | | | | | | | | n/a |
| *Encyonema minutum* (Hilse) Mann | S | | | | | | | | | n/a |
| *Encyonema silesiacum* (Bleisch) Mann | P/S/E | S | S | | | | | | | <1‰ (freshwater) |
| *Encyonema sprechmannii* Metzeltin, Lange-Bertalot & García-Rodríguez | S | | | | | | | | | n/a |
| *Encyonopsis microcephala* (Grunow) Krammer | S | | | | | | | | | |
| *Entomoneis alata* (Ehrenberg) Ehrenberg | | | | | S/P | | | | | 34.5-35.6‰ (marine) |
| *Entopyla australis* (Ehrenberg) Ehrenberg | | | | | | | | | P | 32-34‰ (marine) |
| *Epithemia adnata* (Kützing) Brébisson | | S | S | S | | | | | | <1‰ (freshwater) |
| *Epithemia argus* (Ehrenberg) Kützing | | S | S | | | | | | | 0.5-5‰ (freshwater/brackish) |
| *Epithemia sorex* Kützing | | S | | | | | | P | | 0-10‰ (freshwater/brackish) |
| *Epithemia turgida* var. *granulata* (Ehrenberg) Brun | S | | | | | | | | | n/a |
| *Eunotia arcus* Ehrenberg | P | | | | | | | | | 0-1.2‰ (freshwater) |
| *Eunotia bilunaris* (Ehrenberg) Schaarschmidt | P | | | | | | | | | 0-10.5‰ (freshwater/brackish) |
| *Eunotia biseriata* Hustedt | S | | | | | | | | | n/a |
| *Eunotia camelus* Ehrenberg | S | | | | | | | | | n/a |
| *Eunotia formica* Ehrenberg | P | | | | | | | | | 0-0.2‰ (freshwater) |
| *Eunotia hexaglyphis* Ehrenberg | P | | | | | | | | | 0-0.25‰ (freshwater) |
| *Eunotia implicata* Nörpel, Lange-Bertalot & Alles | S | | | | | | | | | n/a |
| *Eunotia incisa* Smith | S | | | | | | | | | n/a |
| *Eunotia larra* Frenguelli | S | | | | | | | | | n/a |

(Continued)

| TAXA NAME/AUTHORITY | 1 | 2 | 3 | 4 | 5 | 6 | 7 | 8 | 9 | SALINITY |
|---|---|---|---|---|---|---|---|---|---|---|
| *Eunotia luna* var. *aequalis* Hustedt | S | | | | | | | | | n/a |
| *Eunotia major* var. *gigantea* Frenguelli | S | | | | | | | | | n/a |
| *Eunotia major* var. *major* (Schmith) Rabenhorst | S | | | | | | | | | n/a |
| *Eunotia monodon* Ehrenberg | P | | | | | | | | | 0-0.2‰ (freshwater) |
| *Eunotia monodon* var. *bidens* (Gregory) Hustedt | S | | | | | | | | | n/a |
| *Eunotia odebrechtiana* Metzeltin & Lange-Bertalot | S | | | | | | | | | n/a |
| *Eunotia pectinalis* var. *undulata* (Ralfs) Rabenhorst | P | | | | | | | | | 0-0.5‰ (freshwater) |
| *Eunotia pyramidata* var. *monodon* Krasske | S | | | | | | | | | n/a |
| *Eunotia praerupta* Ehrenberg | P | | | | | | | | | 0-0.25‰ (freshwater) |
| *Eunotia praerupta* var. *excelsa* Krasske | P | | | | | | | | | 0.2-0.4‰ (freshwater) |
| *Eunotia tecta* Krasske | S | | | | | | | | | n/a |
| *Eunotia tridentula* Ehrenberg | S | | | | | | | | | n/a |
| *Eunotia veneris* (Kützing) De Toni | S | | | | | | | | | n/a |
| *Fallacia monoculata* (Hustedt) Mann | S | | | | | | | | | n/a |
| *Fallacia omissa* (Hustedt) Mann | S | | | | | | | | | n/a |
| *Fallacia pygmaea* (Kützing) Stickle & Mann | S/E | S | S | S | | | | | | WA: 20.8±6.3‰ (marine/brackish) |
| *Fistulifera saprophila* (Lange-Bertalot &. Bonik) Lange-Bertalot | S | | | | | | | | | n/a |
| *Fragilaria capucina* Desmazières | S/P | | | | | | | | | 0-1.2‰ (freshwater) |
| *Fragilaria capucina* subsp. *rumpens* (Kützing) Lange-Bertalot | S | | | | | | | | | n/a |
| *Fragilaria capucina* var. *vaucheriae* (Kützing) Lange-Bertalot | S | | | | | | | | | n/a |

| TAXA NAME/AUTHORITY | 1 | 2 | 3 | 4 | 5 | 6 | 7 | 8 | 9 | SALINITY |
|---|---|---|---|---|---|---|---|---|---|---|
| *Fragilaria crassa* Metzeltin & Lange-Bertalot | S | | | | | | | | | n/a |
| *Fragilaria crotonensis* Kitton | | | | | | P | | | | 0-26‰ (brackish/marine, euryhaline) |
| *Fragilaria goulardii* (Brébisson) Lange-Bertalot | S | | | | | | | | | n/a |
| *Fragilaria heidenii* Østrup | P/S | | | | | | | | | 0-6.5‰ (freshwater/brackish) |
| *Fragilaria tenera* (Smith) Lange-Bertalot | S | | | | | | | | | n/a |
| *Fragilariforma virescens* (Ralfs) Williams & Round | | S | S | S | | | | | | WA: 6.4±7‰ (freshwater/brackish) |
| *Frankophila similioides* Lange-Bertalot & Rumrich | S | | | | | | | | | n/a |
| *Frustulia neomundana* Lange-Bertalot & Rumrich | S | | | | | | | | | n/a |
| *Frustulia rhomboides* (Ehrenberg) De Toni | | | S | S | | | | | | <1‰ (freshwater) |
| *Frustulia rhomboides* var. *viridula* (Brébisson) Cleve | | | | | | | | | P | 32-34‰ (marine) |
| *Frustulia vulgaris* (Twaites) De Toni | | | | | | | | P | | <1‰ (freshwater) |
| *Geissleria decussis* (Østrup) Lange-Bertalot & Metzeltin | S | | | | | | | | | n/a |
| *Geissleria ignota* (Krasske) Lange-Bertalot & Metzeltin | S | | | | | | | | | n/a |
| *Geissleria perelegans* (Hustedt) Metzeltin & Lange-Bertalot | S | | | | | | | | | n/a |
| *Geissleria schmidtiae* Lange-Bertalot & Rumrich | S | | | | | | | | | n/a |
| *Gomphoneis minuta* (Stone) Kociolek & Stoermer | | | | | | | | P | | <1‰ (freshwater) |
| *Gomphoneis herculeana* (Ehrenberg) Cleve | | | | | | | | P | | 0-35 |
| *Gomphonema* spp. | | | | | | | | P | | - |
| *Gomphonema abbreviatum* (Agardh) Kützing | | S | S | | | | | | | 0.5-20‰ (marine/brackish) |
| *Gomphonema acuminatum* Ehrenberg | | | | | | | | P | | <1‰ (freshwater) |

(Continued)

| TAXA NAME/AUTHORITY | 1 | 2 | 3 | 4 | 5 | 6 | 7 | 8 | 9 | SALINITY |
|---|---|---|---|---|---|---|---|---|---|---|
| *Gomphonema affine* Kützing | S | | | | | | | | | n/a |
| *Gomphonema affine* var. *rhombicum* Reichardt | S | | | | | | | | | n/a |
| *Gomphonema anglicum* Ehrenberg | S | | | | | | | | | n/a |
| *Gomphonema angustatum* (Kützing) Rabenhorst | | S | S | S | | | | | | 8-22‰ (marine/brackish) |
| *Gomphonema apicatum* Ehrenberg | S | | | | | | | | | n/a |
| *Gomphonema augur* Ehrenberg | P/S/E | | | | | | | | | 0-0.2‰ (freshwater) |
| *Gomphonema auritum* Braun | S | | | | | | | | | n/a |
| *Gomphonema capitatum* Ehrenberg | S | | | | | | | | | n/a |
| *Gomphonema clavatum* Ehrenberg | P | | | | | | | | | 0.2-0.4‰ (freshwater) |
| *Gomphonema gracile* Ehrenberg | P | | | | | | | | | 0-0.2‰ (freshwater) |
| *Gomphonema lagenula* Kützing | S | | | | | | | | | n/a |
| *Gomphonema laticollum* Reichardt | S | | | | | | | | | n/a |
| *Gomphonema olivaceum* (Lyngbye) Kützing | | S | S | S | | | | P | | 0-5‰ (freshwater/brackish) |
| *Gomphonema parvulum* (Kützing) Grunow | P/S/E | S | S | S | | | | | | WA: 3±3.7‰ (freshwater/brackish) |
| *Gomphonema pseudotenellum* Lange-Bertalot | | | | | | | | P | | <1‰ (freshwater) |
| *Gomphonema salae* Lange-Bertalot & Reichardt | S | | | | | | | | | n/a |
| *Gomphonema truncatum* Ehrenberg | P | S | S | S | | | | P | | 0-0.1 (freshwater) |
| *Gomphonema turris* Ehrenberg | S | | | | | | | | | n/a |
| *Gomphonema turris* var. *brasiliensis* (Fricke) Frenguelli | S | | | | | | | | | n/a |
| *Grammatophora angulosa* Ehrenberg | | | | | | | | | P | 32-34‰ (marine) |
| *Grammatophora hamulifera* Kützing | | | | | | | | | P | 32-34‰ (marine) |

(Continued)

| TAXA NAME/AUTHORITY | 1 | 2 | 3 | 4 | 5 | 6 | 7 | 8 | 9 | SALINITY |
|---|---|---|---|---|---|---|---|---|---|---|
| *Grammatophora marina* (Lyngbye) Kützing | | | | | | | | P | P | 3-35‰ (marine/brackish) |
| *Grammatophora oceanica* Ehrenberg | | S | S | | | | | | | 32-34‰ (marine) |
| *Grammatophora serpentina* Ehrenberg | | | | | | | | | P | 32-34‰ (marine) |
| *Grammatophora undulata* Ehrenberg | S | | | | | | | | | n/a |
| *Guinardia delicatula* (Cleve) Hasle | | | | | P | | | | | 30-33‰ (marine) |
| *Guinardia flaccida* (Castracane) Peragallo | | | | | P | | | | | 30-33‰ (marine) |
| *Gyrosigma* spp. | | S | S | S | | P | | P | | - |
| *Gyrosigma acuminatum* (Kützing) Rabenhorst | | | | | | | | P | | <1‰ (freshwater) |
| *Gyrosigma attenuata* (Kützing) Rabenhorst | P | | | | S/P | P | | | | 0-35‰ (euryhaline) |
| *Gyrosigma fasciola* (Ehrenberg) Griffith & Henfrey | | | | | S | | | | | 24-26‰ (brackish/marine) |
| *Gyrosigma scalproides* (Rabenhorst) Cleve | P | | | | | | | | | 0-0.2‰ (freshwater) |
| *Gyrosigma spencerii* (Bailey) Griffith & Henfrey | P | | | | | | | | | 0-15‰ (freshwater/brackish) |
| *Hantzschia amphioxys* (Ehrenberg) Grunow | P/S/E | | S | S | | | | | | 0-0.2‰ (freshwater) |
| *Hantzschia amphioxys* var. *capitellata* | S | | | | | | | | | n/a |
| *Hantzschia uruguayensis* Metzeltin, Lange-Bertalot & García-Rodríguez | S | | | | | | | | | n/a |
| *Hantzschia virgata* var. *capitellata* Hustedt | | | | S | | | | | | WA: 19.3±3‰ (brackish/marine) |
| *Hantzschia vivax* (Smith) Tempère | S | | | | | | | | | n/a |
| *Helicotheca tamesis* (Shrubsole) Ricard | | | | | | P | | | | 26‰ (marine/brackish) |
| *Hemiaulus sinensis* Greville | | | | | P | | | | | n/a |
| *Hippodonta capitata* (Ehr.) Lange-Bert., Metz. & Witk. | P/S/E | | | | | | | | | 0-0.2‰ (freshwater) |

(Continued)

| TAXA NAME/AUTHORITY | 1 | 2 | 3 | 4 | 5 | 6 | 7 | 8 | 9 | SALINITY |
|---|---|---|---|---|---|---|---|---|---|---|
| *Hippodonta hungarica* (Grun.) Lange-Bert., Metz. & Witk. | P/S | S | S | S | | | | | | WA: 4.9±6.7‰ (freshwater/brackish) |
| *Hippodonta linearis* (Østrup) Lange-Bert, Metz & Witk. | | S | S | | | | | | | 7-22‰ (marine/brackish) |
| *Hippodonta luneburgensis* (Grun.) Lange-Bert., Metz. & Witk. | | S | S | | | | | | | 7-22‰ (marine/brackish) |
| *Hippodonta subtilissima* Lange-Bertalot | S | | | | | | | | | n/a |
| *Hyalodiscus radiatus* (O' Meara) Grunow | | | | | | | | | P | 32-34‰ (marine) |
| *Hyalodiscus scoticus* (Kützing) Grunow | | | | | | | | | P | 32-34‰ (marine) |
| *Hyalodiscus subtilis* Bailey | | S | S | S | | | | | P | 32-34‰ (marine) |
| *Karayevia clevei* (Grunow) Round & Bukhtiyarova | S | | | | | | | | | n/a |
| *Lemnicola hungarica* (Grunow) Round & Basson | S/E | | | | | | | | | <1‰ (freshwater) |
| *Leptocylindrus* sp. | | | | | | P | | | | - |
| *Licmophora* sp. | | | | | | | | P | | - |
| *Licmophora abbreviata* Agardh | | | | | | | | | P | 32-34‰ (marine) |
| *Licmophora flabellata* Agardh | | | | | | | | | P | 32-34‰ (marine) |
| *Lithodesmium undulatum* Ehrenberg | | | | | P | | | P | | n/a |
| *Luticola charcotii* var. *magelanica* (Hustedt) Metzeltin | S | | | | | | | | | n/a |
| *Luticola claudiae* Metzeltin, Lange-Bertalot & García-Rodriguez | S | | | | | | | | | n/a |
| *Luticola cohnii* (Hilse) Mann | S/E | | | | | | | | | <1‰ (freshwater) |
| *Luticola dapalis* (Frenguelli) Mann | S | | | | | | | | | n/a |
| *Luticola dapaloides* (Frenguelli) Metzeltin & Lange-Bertalot | S | | | | | | | | | n/a |

(Continued)

| TAXA NAME/AUTHORITY | 1 | 2 | 3 | 4 | 5 | 6 | 7 | 8 | 9 | SALINITY |
|---|---|---|---|---|---|---|---|---|---|---|
| *Luticola frenguellii* Metzeltin & Lange-Bertalot | S | | | | | | | | | n/a |
| *Luticola goeppertiana* (Bleisch) Mann | S/E | | | | | | | | | <1‰ (freshwater) |
| *Luticola mutica* (Kützing) Mann | S | S | S | S | | | | P | | 1-7‰ (brackish/freshwater) |
| *Luticola nivalis* (Ehrenberg) Mann | S | | | | | | | | | n/a |
| *Luticola ventricosa* (Kützing) Mann | S/E | | | | | | | | | <1‰ (freshwater) |
| *Luticola saxophila* (Bock) Mann | S | | | | | | | | | n/a |
| *Luticola undulata* (Hilse) Mann | S | | | | | | | | | n/a |
| *Luticola undulata* var. *chilensis* (Hustedt) Metzeltin | S | | | | | | | | | n/a |
| *Lyrella david-mannii* Witkowski, Lange-Bertalot & Metzeltin | S | | | | | | | | | n/a |
| *Lyrella lyra* (Ehrenberg) Karajeva | | | | | | | | | P | 32-34‰ (marine) |
| *Mastogloia belaensis* Voigt | | S | S | | | | | | | 3-9‰ (brackish) |
| *Mastogloia elliptica* (Agardh) Cleve | | S | S | S | | | | | | 3-28‰ (marine/brackish, euryhaline) |
| *Mayamea atomus* (Kützing) Lange-Bertalot | P | | | | | | | | | 0-0.25‰ (freshwater) |
| *Melosira* sp. | | | | | | P | | | | - |
| *Melosira fausta* Schmidt | | | | | | | | | P | 32-34‰ (marine) |
| *Melosira moniliformis* (Müller) Agardh | | | | | P | | | | | n/a |
| *Melosira moniliformis* var. *octagona* (Grunow) Hustedt | S | | | | | | | | | n/a |
| *Melosira nummuloides* Agardh | | | | | | | | | P | 32-34‰ (marine) |
| *Melosira varians* Agardh | S | S | S | S | | P | | P | | <1‰ (freshwater) |
| *Navicella pusilla* (Grunow) Krammer | S | S | S | | | | | | | WA: 6.5±9.5‰ (freshwater/brackish) |

| TAXA NAME/AUTHORITY | 1 | 2 | 3 | 4 | 5 | 6 | 7 | 8 | 9 | SALINITY |
|---|---|---|---|---|---|---|---|---|---|---|
| (Continued) | | | | | | | | | | |
| *Navicula* spp. | | | | | S/P | P | | P | | - |
| *Navicula angusta* Grunow | S | | | | | | | | | n/a |
| *Navicula antonii* Lange-Bertalot | S | | | | | | | | | n/a |
| *Navicula atomus* (Kützing) Grunow | S | | | | | | | | | n/a |
| *Navicula breitenbuchii* Lange-Bertalot | S | | | | | | | | | n/a |
| *Navicula capitatoradiata* Germain | | | | | | | | P | | n/a |
| *Navicula caterva* Hohn & Hellermann | | | S | | | | | | | 2‰ (brackish) |
| *Navicula cincta* (Ehrenberg) Kützing | | S | S | S | | | | | | WA: 12±11.6‰ (brackish/marine) |
| *Navicula constans* Hustedt | P | | | | | | | | | 0-0.25‰ (freshwater) |
| *Navicula cryptocephala* Kützing | P | S | S | | | | | | | 0-6‰ (brackish/freshwater) |
| *Navicula cryptotenella* Lange-Bertalot | S | | | | | | | | | n/a |
| *Navicula cryptotenelloides* Lange-Bertalot | S | | | | | | | | | n/a |
| *Navicula digitatoradiata* (Gregory) Ralfs | | S | | | | | | | | 2.5-8.5‰ (brackish) |
| *Navicula eichhorniaephila* Manguin | S | | | | | | | | | n/a |
| *Navicula elmorei* Patrick | P | | | | | | | | | 0-0.3‰ (freshwater) |
| *Navicula endophytica* Hasle | | S | S | | | | | | | <1‰ (freshwater) |
| *Navicula erifuga* Lange-Bertalot | P/S/E | | | | | | | | | 0-0.4‰ (freshwater) |
| *Navicula exigua* Gregory | P | | | | | | | | | 0-0.3‰ (freshwater) |
| *Navicula forcipata* var. *densestriata* Schmidt | | | | | | | | | P | 32-34‰ (marine) |
| *Navicula gregaria* Donkin | S/E | S | S | S | | | | P | | WA: 12±10.5‰ (brackish/marine) |
| *Navicula lanceolata* var. *arenaria* (Donkin) Van Heurck | | S | S | S | | | | | | 8-28‰ (marine/brackish) |

(Continued)

| TAXA NAME/AUTHORITY | 1 | 2 | 3 | 4 | 5 | 6 | 7 | 8 | 9 | SALINITY |
|---|---|---|---|---|---|---|---|---|---|---|
| *Navicula laterostrata* Hustedt | S | | | | | | | | | n/a |
| *Navicula longicephala* Hustedt | S | | | | | | | | | n/a |
| *Navicula microcari* Lange-Bertalot | S | | | | | | | | | n/a |
| *Navicula notha* Wallace | P | | | | | | | | | 0-0.3‰ (freshwater) |
| *Navicula novaesiberica* Lange-Bertalot | S | | | | | | | | | n/a |
| *Navicula peregrina* (Ehrenberg) Kützing | P | S | S | S | | | | P | | 0-6.5‰ (brackish/freshwater) |
| *Navicula peregrinopsis* Lange-Bertalot & Witkowski | S | | | | | | | | | n/a |
| *Navicula pseudotenelloides* Krasske | S | | | | | | | | | n/a |
| *Navicula radiosa* Kützing | | | | | | | | P | | 0-10‰ (brackish/freshwater) |
| *Navicula rhynchocephala* Kützing | P/S/E | | | | | | | | | 0-0.2‰ (freshwater) |
| *Navicula rostellata* Kützing | S | | | | | | | | | n/a |
| *Navicula sanctaecrucis* Østrup | S | | | | | | | | | n/a |
| *Navicula schroeteri* Meister | S | | | | | | | | | n/a |
| *Navicula symmetrica* Patrick | S | | | | | | | | | n/a |
| *Navicula tackei* f. *major* Maidana & Herbst | | | | | | | | P | | <1‰ (freshwater) |
| *Navicula tenelloides* Hustedt | S | | | | | | | | | <1‰ (freshwater) |
| *Navicula tripunctata* (Müller) Bory | | S | S | S | | | | P | | 0-2‰ (freshwater/brackish) |
| *Navicula trivialis* Lange-Bertalot | P/S/E | S | | | | | | | | 0-6.5‰ (brackish/freshwater) |
| *Navicula veneta* Kützing | S/E | | | | | | | P | | <1‰ (freshwater) |
| *Neidium affine* (Ehrenberg) Pfitzer | P/S/E | | | | | | | | | 0-0.4‰ (freshwater) |
| *Neidium affine* var. *longiceps* (Gregory) Cleve | S | | | | | | | | | n/a |

| TAXA NAME/AUTHORITY | 1 | 2 | 3 | 4 | 5 | 6 | 7 | 8 | 9 | SALINITY |
|---|---|---|---|---|---|---|---|---|---|---|
| *Neidium amphirhynchus* (Ehrenberg) Pfitzer | S | | | | | | | | | n/a |
| *Neidium ampliatum* (Ehrenberg) Krammer | S | | | | | | | | | n/a |
| *Neidium catarinense* (Krasske) Lange-Bertalot | S | | | | | | | | | n/a |
| *Neidium dubium* (Ehenberg) Cleve | S/E | | | | | | | | | <1‰ (freshwater) |
| *Neidium hercynicum* Mayer | S | | | | | | | | | n/a |
| *Neidium iridis* (Ehrenberg) Cleve | S | | | | | | | | | n/a |
| *Neidium iridis var. amphigomphus* (Ehrenberg) Tempere & Peragallo | S | | | | | | | | | n/a |
| *Neidium iridis var. intercedens* Mayer | P | | | | | | | | | 0-0.4‰ (freshwater) |
| *Neidium magellanica var. minor* Frenguelli | S | | | | | | | | | n/a |
| *Neocalyptrella robusta* (Norman) Hern-Bec. & Meave | | | | | | | | | P | 32-34‰ (marine) |
| *Nitzschia* spp. | | S | | S | S | | | P | | - |
| *Nitzschia acicularis* (Kützing) Smith | P/S/E | | | | | | | | | 0-0.4‰ (freshwater) |
| *Nitzschia amphibia* Grunow | S | S | S | S | | | | | | WA: 3.2±4.6‰ (brackish/freshwater) |
| *Nitzschia angularis* Smith | | | | | | | | | P | 32-34‰ (marine) |
| *Nitzschia brevissima* Grunow | P/S/E | | | | | | | | | 0.2-0.4‰ (freshwater) |
| *Nitzschia capitellata* Hustedt | S | | | | | | | | | n/a |
| *Nitzschia clausii* Hantzsch | S/E | S | S | S | | | | | | 0-7‰ (brackish/freshwater) |
| *Nitzschia commutata* Grunow | S | | | | | | | | | n/a |
| *Nitzschia commutatoides* Lange-Bertalot | P | | | | | | | | | 0-0.4‰ (freshwater) |
| *Nitzschia constricta* (Gregory) Grunow | P | | | | | | | P | P | 0-34‰ (marine/brackish, euryhaline) |
| *Nitzschia draveillensis* Coste & Ricard | P/S/E | | | | | | | | | 0-0.4‰ (freshwater) |

| TAXA NAME/AUTHORITY | 1 | 2 | 3 | 4 | 5 | 6 | 7 | 8 | 9 | SALINITY |
|---|---|---|---|---|---|---|---|---|---|---|
| *Nitzschia filiformis* (Smith) Hustedt | P/S/E | | | | | | | | | 0-0.5‰ (freshwater) |
| *Nitzschia filiformis* var. *conferta* (Richt) Lange-Bertalot | S | | | | | | | | | n/a |
| *Nitzschia fonticola* (Grunow) Grunow | S/E | | | | | | | | | <1‰ (freshwater) |
| *Nitzschia frustulum* (Kützing) Grunow | P/S/E | | | | | | | | | 0-7‰ (brackish/freshwater) |
| *Nitzschia fruticosa* Hustedt | P | | | | | | | | | 0-0.2‰ (freshwater) |
| *Nitzschia gracilis* Hantzsch | P | | | | | | | | | 0-0.5‰ (freshwater) |
| *Nitzschia habirshawii* Febiger | | | | | | | | | P | 32-34‰ |
| *Nitzschia hantzschiana* Rabenhorst | | S | | | | | | | | n/a |
| *Nitzschia heidenii* (Meister) Hustedt | S | | | | | | | | | n/a |
| *Nitzschia inconspicua* Grunow | S | S | S | S | | | | | | WA: 6.4±7.6‰ (brackish/freshwater) |
| *Nitzschia lacunarum* Hustedt | S | | | | | | | | | n/a |
| *Nitzschia linearis* (Agardh) Smith | P/S/E | | | | | | | | | 0-0.2‰ (freshwater) |
| *Nitzschia lorenziana* Grunow | S | | | | | | | | | n/a |
| *Nitzschia microcephala* Grunow | | S | S | S | | | | | | WA: 7.6±5.5‰ (brackish/freshwater) |
| *Nitzschia nana* Grunow | P/S/E | S | | | | | | | | 0-0.2‰ (freshwater) |
| *Nitzschia palea* (Kützing) Smith | P/S/E | | | | | | | | | 0-0.2‰ (freshwater) |
| *Nitzschia paleacea* Grunow | P/S | | | | | | | | | 0-0.2‰ (freshwater) |
| *Nitzschia perminutum* (Grunow) Peragallo | S | | | | | | | | | n/a |
| *Nitzschia pumila* Hustedt | S | | | | | | | | | n/a |
| *Nitzschia rautenbachiae* Cholnoky | | S | S | S | | | | | | WA: 10±5‰ (brackish) |
| *Nitzschia reversa* Smith | S | | | | | | | | | n/a |

(Continued)

| TAXA NAME/AUTHORITY | 1 | 2 | 3 | 4 | 5 | 6 | 7 | 8 | 9 | SALINITY |
|---|---|---|---|---|---|---|---|---|---|---|
| *Nitzschia scalpelliformis* Grunow | S | | | | | | | | | n/a |
| *Nitzschia sigma* (Kützing) Smith | P/S | S | | S | S | | | | | 0-34.5‰ (marine/brackish, euryhaline) |
| *Nitzschia sigmoidea* (Nitzsch) Smith | S | | | | | | | | | n/a |
| *Nitzschia sinuata* var. *delongei* (Grunow) Lange-Bertalot | S | | | | | | | | | n/a |
| *Nitzschia socialis* Gregory | | | | S | | | | | | 29‰ (marine) |
| *Nitzschia subconstricta* Grunow | S | | | | | | | | | n/a |
| *Nitzschia umbonata* (Ehrenberg) Lange-Bertalot | S | | | | | | | | | n/a |
| *Nitzschia vermicularis* (Kützing) Hantzsch | P | | | | | | | | | 0-0.4‰ (freshwater) |
| *Nitzschia vitrea* Norman | | S | S | | | | | | | 2-20‰ (brackish/marine) |
| *Nupela lesothensis* (Schoeman) Lange-Bertalot | S | | | | | | | | | n/a |
| *Odontella* sp. | | | | S | | | | | | - |
| *Odontella aurita* (Lyngbye) Agardh | | | | | | | | P | P | 25-35‰ (marine/brackish) |
| *Odontella mobiliensis* (Bailey) Grunow | | | | | P | | | P | | 25-35‰ (marine/brackish) |
| *Odontella obtusa* Kütz. | | | | | | | | | P | 32-34‰ (marine) |
| *Odontella sinensis* (Greville) Grunow | | | | | P | | | | | n/a |
| *Opephora* sp. | | S | S | | | | | | | - |
| *Opephora marina* (Gregory) Petit | | | | | | | S | | | 38‰ (marine) |
| *Opephora pacifica* (Grunow) Petit | | S | S | S | | | | | | WA: 13±6‰ (brackish/marine) |
| *Orthoseira roeseana* (Rabenhorst) O'Meara | S | | | | | | | | | n/a |
| *Paralia sulcata* (Ehrenberg) Cleve | | S | S | S | S/P | P | | P | P | WA: 26±3‰ /0-35‰ (marine/brackish, euryhaline) |

| TAXA NAME/AUTHORITY | 1 | 2 | 3 | 4 | 5 | 6 | 7 | 8 | 9 | SALINITY |
|---|---|---|---|---|---|---|---|---|---|---|
| *Petrodictyon gemma* (Ehrenberg) Mann | | | | | S | | | | | 30-36‰ (marine) |
| *Petroneis monilifera* (Cleve) Stickle & Mann | | S | | | | | | | | 20-25‰ (marine/brackish) |
| *Pinnularia acrosphaeria* (Brébisson) Smith | S | | | | | | | | | n/a |
| *Pinnularia acrosphaeria* f. *maxima* Cleve | S | | | | | | | | | n/a |
| *Pinnularia borealis* Ehrenberg | S | S | S | S | | | | P | P | 3-20‰ (brackish/marine, euryhaline) |
| *Pinnularia borealis* var. *islandica* Krammer | S | | | | | | | | | n/a |
| *Pinnularia borealis* var. *scalaris* (Ehrenberg) Rabenhorst | S | | | | | | | | | n/a |
| *Pinnularia borealis* var. *sublinearis* Krammer | S | | | | | | | | | n/a |
| *Pinnularia brevicostata* Cleve | | | | | | | | P | | <1‰ (freshwater) |
| *Pinnularia carambolae* Frenguelli | S | | | | | | | | | n/a |
| *Pinnularia divergens* var. *elliptica* Grunow | S | | | | | | | | | n/a |
| *Pinnularia divergens* var. *malayensis* Hustedt | S | | | | | | | | | n/a |
| *Pinnularia divergens* var. *sublinearis* Cleve | S | | | | | | | | | n/a |
| *Pinnularia divergens* var. *undulata* Peragallo & Héribaud | S | | | | | | | | | n/a |
| *Pinnularia divergens* var. *protracta* Krammer, & Metzeltin | S | | | | | | | | | n/a |
| *Pinnularia doehringii* Frenguelli | S | | | | | | | | | n/a |
| *Pinnularia dubitabilis* Hustedt | S | | | | | | | | | n/a |
| *Pinnularia ehrlichiana* Metzeltin, Lange-Bertalot & Garcia-Rodriguez | S | | | | | | | | | n/a |
| *Pinnularia fistuciformis* Metzeltin, Lange-Bertalot & Garcia-Rodriguez | S | | | | | | | | | n/a |
| *Pinnularia gibba* Ehrenberg | S/E | | | | | | | | | <1‰ (freshwater) |

| TAXA NAME/AUTHORITY | 1 | 2 | 3 | 4 | 5 | 6 | 7 | 8 | 9 | SALINITY |
|---|---|---|---|---|---|---|---|---|---|---|
| *Pinnularia hyalina* Hustedt | S | | | | | | | | | n/a |
| *Pinnularia* aff. *joculata* (Manguin) Krammer | S | | | | | | | | | n/a |
| *Pinnularia latevittata* Cleve | S | | | | | | | | | n/a |
| *Pinnularia maior* (Kützing) Cleve | P | | | | | | | | | 0-0.4‰ (freshwater) |
| *Pinnularia marchica* Ilka Schönfelder | S | | | | | | | | | n/a |
| *Pinnularia mesolepta* (Ehrenberg) Smith | P | | | | | | | | | 0-0.5‰ (freshwater) |
| *Pinnularia microstauron* (Ehrenberg) Cleve | P | | | | | | | | P | 0-0.5‰ (freshwater) |
| *Pinnularia neomajor* Krammer | S | | | | | | | | | n/a |
| *Pinnularia neuquina* Frenguelli | S | | | | | | | | | n/a |
| *Pinnularia nitzschiophila* Rumrich | S | | | | | | | | | n/a |
| *Pinnularia rabenhorstii* var. *franconia* Krammer | S | | | | | | | | | n/a |
| *Pinnularia schweinfurthii* (Schmidt) Patrick | S | | | | | | | | | n/a |
| *Pinnularia subacoricola* Metzeltin, Lange-Bertalot & Garcia-Rodriguez | S | | | | | | | | | n/a |
| *Pinnularia subanglica* Krammer | S | | | | | | | | | n/a |
| *Pinnularia* cf. *subcapitata* Gregory | S/E | | | | | | | | | <1‰ (freshwater) |
| *Pinnularia* spec. cf. *stomatophora* var. *salina* Krammer | S | | | | | | | | | n/a |
| *Pinnularia tabellaria* Ehrenberg | S | | | | | | | | | n/a |
| *Pinnularia viridiformis* Krammer | S | | | | | | | | | n/a |
| *Pinnularia viridis* (Nitzsch) Ehrenberg | | S | | | | | | | | 20‰ (brackish) |
| *Placoneis clementis* (Grunow) Cox | S/E | | | | | | | | | <1‰ (freshwater) |

(Continued)

| TAXA NAME/AUTHORITY | 1 | 2 | 3 | 4 | 5 | 6 | 7 | 8 | 9 | SALINITY |
|---|---|---|---|---|---|---|---|---|---|---|
| *Placoneis disparilis* (Hustedt) Metzeltin & Lange-Bertalot | S | | | | | | | | | n/a |
| *Placoneis gastrum* (Ehrenberg) Mereschkovsky | S | S | | | | | | | | 14‰ (brackish) |
| *Placoneis placentula* (Ehrenberg) Mereschkowsky | P/S | | | | | | | | | 0-0.3‰ (freshwater) |
| *Placoneis parelginensis* (Gregory) Cox | S | | | | | | | | | n/a |
| *Placoneis serena* (Frenguelli) Metzeltin | S | | | | | | | | | n/a |
| *Plagiogramma staurophorum* (Gregory) Heiberg | | | S | | | | | | P | WA: 22±6‰ (marine/brackish) |
| *Planothidium delicatulum* (Kützing) Round & Bukht. | S/E | S | S | S | | | S | | | WA: 6.7±6.7 (brackish) |
| *Planothidium lanceolatum* (Brébisson) Lange-Bert. | | S | S | S | | | S | | | 0-38‰ (marine/brackish, euryhaline) |
| *Pleurosigma* spp. | | S | | S | | | | P | | - |
| *Pleurosigma angulatum* Smith | | | | | | P | | | | 25-35‰ (marine/brackish) |
| *Pleurosigma elongatum* Smith | P | | | | | | | | | 18‰ (marine/brackish) |
| *Pleurosigma normanii* Ralfs | P | | | | | | | | P | 32-34‰ (marine) |
| *Pleurosigma strigosum* Smith | | | | | | | | | P | 32-34‰ (marine) |
| *Pleurosira laevis* (Ehrenberg) Compère | S/E | S | S | S | | | | P | | WA: 4±4.7‰ (brackish/freshwater) |
| *Podosira* sp. | | | | | | P | | | | - |
| *Podosira maxima* (Kützing) Grunow | | | | | | | | | P | 32-34‰ (marine) |
| *Podosira montagnei* Kützing | | | | | | | | | P | 32-34‰ (marine) |
| *Podosira stelligera* (Bailey) Mann | | S | S | S | P | | S | | | 33-38‰ (marine) |
| *Psammodictyon constrictum* (Gregory) Mann | | S | S | S | | | | | | WA: 10.5±6‰ (brackish/marine) |
| *Psammodictyon panduriforme* (Gregory) Mann | | S | S | S | | | | | | 20-28‰ (marine/brackish) |
| *Pseudo-nitzschia* spp. | P | | | | | | | | | - |

| TAXA NAME/AUTHORITY | 1 | 2 | 3 | 4 | 5 | 6 | 7 | 8 | 9 | SALINITY |
|---|---|---|---|---|---|---|---|---|---|---|
| *Pseudo-nitzschia seriata* (Cleve) Peragallo | | | | | P | | | | | n/a |
| *Pseudostaurosira brevistriata* (Grunow) Williams & Round | S | S | S | S | | | | | | WA: 7.8±6.75‰ (brackish/freshwater) |
| *Reimeria sinuata* (Gregory) Kociolek & Stoermer | S | S | S | S | | | | P | | 0-5‰ (freshwater/brackish) |
| *Reimeria uniseriata* Sala, Guerrero & Ferrario | S | | | | | | | | | n/a |
| *Rhabdonema adriaticum* Kützing | | | | | | | | P | P | 30-35‰ (marine) |
| *Rhabdonema arcuatum* (Lyngbye) Kützing | S | | | | | | | | P | 32-34‰ (marine) |
| *Rhabdonema minutum* Kützing | | | | | | | | | P | 32-34‰ (marine) |
| *Rhaphoneis amphiceros* (Ehrenberg) Ehrenberg | S | S | S | S | P | | S | P | P | 15-38‰ (marine/brackish) |
| *Rhizosolenia* sp. | P | | | | | P | | | | - |
| *Rhizosolenia hebetata* Bailey | | | | | | | | | P | 32-34‰ (marine) |
| *Rhizosolenia setigera* Brightwell | P | | | | | P | | | | 25-35‰ (marine/brackish) |
| *Rhizosolenia styliformis* Brightwell | | | | | | | | | P | 32-34‰ (marine) |
| *Rhoicosphenia abbreviata* (Agardh) Lange-Bertalot | S | S | S | S | | | | P | | WA: 5.25±6.3‰ (brackish/freshwater) |
| *Rhopalodia brebissonii* Krammer | S/P | S | S | S | | | | | | 0-2‰ (freshwater/brackish) |
| *Rhopalodia gibba* (Ehrenberg) Müller | S | | S | S | | | | P | | 0-0.5‰ (freshwater) |
| *Rhopalodia gibberula* (Ehrenberg) Müller | S | | S | S | | | S | | | WA: 5.3±5.8‰ (brackish/freshwater) |
| *Rhopalodia musculus* (Kützing) Müller | | S | S | S | | | | | | 0-10‰ (brackish/freshwater) |
| *Rhopalodia operculata* (Agardh) Håk. | S | | | | | | | | | n/a |
| *Scoliopleura* sp. | | | | | S | | | | | - |
| *Sellaphora laevissima* (Kützing) Mann | S | | | | | | | | | n/a |

(Continued)

| TAXA NAME/AUTHORITY | 1 | 2 | 3 | 4 | 5 | 6 | 7 | 8 | 9 | SALINITY |
|---|---|---|---|---|---|---|---|---|---|---|
| *Sellaphora nyassensis* (Müller) Mann | P/S | | | | | | | | | 0-0.3‰ (freshwater) |
| *Sellaphora pupula* (Kützing) Mereschkovsky | P/S/E | S | | S | | | | | | 0-0.4‰ (freshwater) |
| *Sellaphora rectangularis* (Gregory) Lange-Bertalot & Metzeltin | S | | | | | | | | | n/a |
| *Sellaphora seminulum* (Grunov) Mann | | S | S | | | | | | | <1‰ (freshwater) |
| *Skeletonema costatum* (Greville) Cleve | P | | | | P | | | | P | 2-34‰ (marine/brackish, euryhaline) |
| *Skeletonema subsalsum* (Cleve) Bethge | S/E | | | | | | | | | <1‰ (freshwater) |
| *Stauroneis* spp. | | S | | | S | | | | | - |
| *Stauroneis anceps* Ehrenberg | S | | | | | | | | | n/a |
| *Stauroneis brasiliensis* (Zimmermann) Compère | S | | | | | | | | | n/a |
| *Stauroneis* cf. *javanica* (Grunov) Cleve | S | | | | | | | | | n/a |
| *Stauroneis obtusa* Lagerstedt | S | | | | | | | | | n/a |
| *Stauroneis phoenicenteron* (Nitzsch) Ehrenberg | S | | | | | | | | | n/a |
| *Stauroneis producta* Grunow | | S | S | S | | | | | | <1‰ (freshwater) |
| *Stauroneis schinzii* var. *maxima* Frenguelli | S | | | | | | | | | n/a |
| *Stauroneis* cf. *schroederi* Hustedt | S | | | | | | | | | n/a |
| *Stauroneis subgracilis* Lange-Bertalot & Krammer | S | | | | | | | | | n/a |
| *Stauroneis tackei* (Hustedt) Krammer & Lange-Bertalot | | S | S | S | | | | | | 0-2.5‰ (freshwater/brackish) |
| *Staurosira altiplanensis* Lange-Bertalot & Rumrich | S | | | | | | | | | n/a |
| *Staurosira construens* Ehrenberg | P | S | S | | | | | | | 0-0.4‰ (freshwater) |
| *Staurosira elliptica* (Schumann) Williams & Round | | S | S | | | | | | | WA: 2.7±4.5‰ (brackish/freshwater) |

| TAXA NAME/AUTHORITY | 1 | 2 | 3 | 4 | 5 | 6 | 7 | 8 | 9 | SALINITY |
|---|---|---|---|---|---|---|---|---|---|---|
| *Staurosira fernandae* García-Rodríguez, Lange-Bertalot & Metzeltin | S | | | | | | | | | n/a |
| *Staurosira* cf. *leptostauron* (Ehrenberg) Hustedt | S | | | | | | | | | n/a |
| *Staurosira longirostris* Frenguelli | S | | | | | | | | | n/a |
| *Staurosira martyi* (Hérib.) Lange-Bertalot | S | | | | | | | | | |
| *Staurosira venter* (Ehrenberg) Kobayasi | | S | S | S | | | | | | WA: 9.7±8.3‰ (brackish/freshwater) |
| *Staurosirella pinnata* (Ehrenberg) Williams & Round | S | S | S | S | | | | | | WA: 6.3±6.4‰ (brackish/freshwater) |
| *Stellarima stellaris* (Roper) Hasle & Sims | | | | | P | | | | | n/a |
| *Stephanodiscus* spp. | P | | | | | | | P | | - |
| *Stephanodiscus hantzschii* Grunow | P/S | | | | | | | | | 0-0.4‰ (freshwater) |
| *Stephanodiscus parvus* Stoermer & Håkansson | P | | | | | | | | | 0-6‰ (freshwater/brackish) |
| *Surirella* spp. | | S | S | | | | | P | | - |
| *Surirella angusta* Kützing | S | | | | | | | | | n/a |
| *Surirella biseriata* Brébisson | S | | | | | | | | | n/a |
| *Surirella brebissonii* Krammer & Lange-Bertalot | S | | | S | | | | | | n/a |
| *Surirella guatimalensis* Ehrenberg | S | | | | | | | | | n/a |
| *Surirella inducta* Schmidt | | S | S | S | | | | | | 2-10‰ (brackish) |
| *Surirella minuta* Brébisson | S | S | S | S | | | | | | 0-2.5‰ (freshwater/brackish) |
| *Surirella minuta* var. *pedaliformis* Frenguelli | S | | | | | | | | | n/a |
| *Surirella ovalis* Brébisson | S/P | S | S | S | | | | | | 0-6‰ (freshwater/brackish) |
| *Surirella ovalis* var. *apiculata* Müller | | S | S | | | | | | | <1‰ (freshwater) |

(Continued)

| TAXA NAME/AUTHORITY | 1 | 2 | 3 | 4 | 5 | 6 | 7 | 8 | 9 | SALINITY |
|---|---|---|---|---|---|---|---|---|---|---|
| *Surirella splendida* (Ehrenberg) Kützing | S | | | | | | | P | | n/a |
| *Surirella striatula* Turpin | S | S | S | S | | | | | | 0.5-10‰ (brackish) |
| *Synedra* sp. | | | | | | | | P | | - |
| *Synedra fulgens* (Greville) Smith | | | | | | | | | P | 32-34‰ (marine) |
| *Synedra platensis* Frenguelli | | S | S | S | | | | | | 0-3‰ (freshwater/brackish) |
| *Synedra tortuosa* Williams & Metzeltin | S | | | | | | | | | n/a |
| *Synedra ulna* var. *claviceps* Hustedt | S | | | | | | | | | n/a |
| *Tabularia investiens* (Smith) Williams & Round | S | | | | | | | | | n/a |
| *Tabularia tabulata* (Agardh) Snoeijs | | S | S | S | | | | | P | 5-34‰ (marine/brackish, euryhaline) |
| *Terpsinoe americana* (Bailey) Ralfs | S | | | | | | | | | n/a |
| *Terpsinoe musica* Ehrenberg | S | | | | | | | | | n/a |
| *Thalassionema nitzschioides* (Grunow) Mereschkowsky | P | | | | P | | | | | 25-35‰ (marine/brackish) |
| *Thalassiosira* spp. | P | | | | P | P | | P | | - |
| *Thalassiosira anguste-lineata* (Schmidt) Fryxell & Hasle | P | | | | P | | | P | | 20-33‰ (marine/brackish) |
| *Thalassiosira curviseriata* Takano | | | | | P | | | | | 25-35‰ (marine/brackish) |
| *Thalassiosira decipiens* (Grunow) Jorgensen | | S | S | S | | | | | | WA: 21±7‰ (marine/brackish) |
| *Thalassiosira eccentrica* (Ehrenberg) Cleve | | S | | S | P | | | P | | 28-33‰ (marine) |
| *Thalassiosira hendeyi* Hasle & Fryxell | | | | | P | | | | | 25-35‰ (marine/brackish) |
| *Thalassiosira hibernalis* Gayoso | | | | | P | | | | | 30-35‰ (marine) |
| *Thalassiosira leptopus* (Grunow) Hasle & Fryxell | | | | | P | | | | | n/a |
| *Thalassiosira minima* Gaarder | | | | | P | | | | | n/a |

(Continued)

| TAXA NAME/AUTHORITY | 1 | 2 | 3 | 4 | 5 | 6 | 7 | 8 | 9 | SALINITY |
|---|---|---|---|---|---|---|---|---|---|---|
| *Thalassiosira pacifica* Gran & Angst | | | | | P | | | | | n/a |
| *Thalassiosira rotula* Meunier | P | | | | P | | | | | 18-33‰ (marine/brackish) |
| *Thalassiosira simonensii* Hasle & Fryxell | | | | | | | | P | | n/a |
| *Trachyneis aspera* (Ehrenberg) Cleve | | | | | | | | | P | 32-34‰ (marine) |
| *Trachyneis aspera* var. *perobliqua* Cleve | | | | | | | | | P | 32-34‰ (marine) |
| *Triceratium* sp. | | | | | | P | | | | - |
| *Triceratium favus* Ehrenberg | | S | | S | | | | P | P | 25-35‰ (marine/brackish) |
| *Tryblionella acuminata* Smith | S/E | | | | | | | | | 32-34‰ (marine) |
| *Tryblionella angustata* Smith | P/S/E | | | | | | | | | 0-0.2‰ (freshwater) |
| *Tryblionella apiculata* Gregory | S | | | | | | | | | n/a |
| *Tryblionella coarctata* (Grunow) Mann | S | | | | | | | | | n/a |
| *Tryblionella compressa* (Bailey) Poulin | S/E | S | S | S | | | S | | | WA: 14±4‰ (brackish) |
| *Tryblionella debilis* Arnott | S | | | | | | | | | n/a |
| *Tryblionella gracilis* Smith | | S | S | S | | | | | | 0-0.5‰ (freshwater) |
| *Tryblionella granulata* (Grunow) Mann | | S | | | | | | | | WA: 10±4‰ (brackish) |
| *Tryblionella hungarica* (Grunow) Frenguelli | P/S/E | | | | | | | | | 0-0.2‰ (freshwater) |
| *Tryblionella levidensis* Smith | P/S | S | S | | | | S | | | 0-0.4‰ (freshwater) |
| *Tryblionella perversa* (Grunow) Mann | S | | | | | | | | | n/a |
| *Ulnaria delicatissima* var. *angustissima* (Grunow) Aboal & Silva | S | | | | | | | | | n/a |
| *Ulnaria acus* (Kützing) Aboal | S | | | | | | | | | n/a |
| *Ulnaria ulna* (Nitzsch) Compère | S/P | S | S | S | | | | P | | 0-6.5‰ (brackish/freshwater) |
| RICHNESS | 356 | 140 | 122 | 106 | 62 | 19 | 15 | 74 | 88 | |

# INDEX

## A

accuracy, 36, 40
acidity, 48
adaptability, xix
algae, 21
ammonia, 8, 9, 47
amphibia, 83
amplitude, 6
assessment, 1
availability, xvii, 57
averaging, 59, 65

## B

banks, 21
barriers, xx, 32
behavior, 57
biogeography, 41
bioindicators, xviii
biomarkers, 49
biomass, 19, 50, 60
brevis, 69

## C

calibration, xix, 36, 48, 59
cell, 49
changing environment, xix
channels, 18, 19, 27
circulation, 14, 16
classes, 40
classification, xix, 37
climate change, xviii
cluster analysis, 28
codes, xviii, 34
coding, 50
colonization, 60
community, 27
compilation, 45
components, 3
composition, xviii, xix, 1, 2, 3, 7, 8, 10, 11, 12, 13, 15, 17, 19, 21, 22, 24, 26, 27, 30, 31, 38, 42, 48, 49, 53
concentration, 10
conductivity, 9
construction, 14
contamination, 48
covering, xviii, 8
culture, 47
cycling, 49

## D

data analysis, 36
data set, 1, 16, 36, 38, 40, 41
death, 55
definition, 3, 59
density, 49
deposition, 1, 21, 43
deposits, xviii, 34, 51
diet, 51

discharges, 5, 32
dissolved oxygen, 8
distribution, xv, xvii, xviii, xix, 1, 3, 7, 8, 9, 11, 15, 16, 21, 22, 36, 42, 43, 49, 52, 54, 55, 56, 59
diversity, 41
division, 47
dominance, 6, 7, 27, 34, 36
draft, 51
drainage, 6
duration, 11

## E

ecology, xiii, xv, xviii, xix, xx, 7, 19, 43, 47, 49, 50, 55, 57
ecosystem, xviii
effluents, 21
environment, xv, xix, 7, 11, 18, 34, 36, 39, 59
environmental change, xix, 59, 60
environmental characteristics, xv, xvii
environmental conditions, xv, xx, 1, 11
environmental control, 54
environmental factors, 12, 15, 43, 57
erosion, xvii
estuarine systems, xvii, xviii, 54
evolution, 33, 34, 48, 55
exposure, 12

## F

fidelity, 2, 3, 55, 56
fisheries, 6
flood, 11
flooding, 12
fluctuations, 12, 33, 36, 38, 42
fluid, 21
focusing, 6
food, xvii
fossil, xviii, xix, 1, 2, 11, 17, 35, 36, 38, 39, 40, 42, 43, 55

## G

gemma, 20, 31, 86
geology, xviii
gracilis, 84, 93

grades, 8
grazing, 48
groups, xviii, 2, 7, 8, 9, 28, 37, 40
growth, 27, 47
growth rate, 47
guidelines, 2

## H

habitat, 1, 2, 43
harvesting, 7
human activity, 16
hypothesis, xv, 43

## I

identification, xviii
images, 45
indicators, xvii, 1, 8, 48, 49, 51, 53, 59
inferences, xv, xx, 3, 40
integration, 38
interactions, xvii
interval, 34
invertebrates, 48

## L

limitation, 40
living environment, 1

## M

macroalgae, 24, 27
marine environment, 8
marsh, 11, 13, 35, 49, 52, 53, 59
measurement, 47
melting, 21
meta-analysis, 55
meteor, 55
microscopy, 49
mixing, 2
models, xix
mollusks, 2
morphology, 12

## N

nematode, 58

nitrates, 8, 9
nutrients, 8, 22

## O

observations, xix, 28, 49
obstruction, 16
omission, 42
orientation, 28

## P

partial least squares regression, 59
particles, 21
phosphates, 8
photosynthesis, 47
phytoplankton, 7, 8, 11, 18, 19, 27, 49, 53, 54, 57
plants, 18, 21
PLS, 36, 59
pollution, 8, 54
poor, 19
population, 2, 47
precipitation, 21
production, xvii, 49, 50
pulse, 34

## Q

quantitative technique, 36

## R

range, 16, 28, 29, 32, 34, 39, 42
recognition, 2, 39
reconstruction, 2, 3, 13, 36, 38, 48, 51
regression, 36
relationship, xv, 9, 15, 33, 36, 49, 50
resistance, xvii
resolution, 55
runoff, 18

## S

salinity, xvii, xviii, xix, 3, 8, 10, 12, 14, 15, 16, 18, 19, 22, 23, 27, 28, 29, 33, 35, 36, 37, 38, 39, 42, 43, 55, 60, 65
salt, 2, 6, 38, 49, 53, 59
sample, 7, 10, 11, 27, 30, 32, 40, 41, 42, 43, 63, 65
sampling, 2, 7, 8, 9, 11, 12, 14, 17, 20, 22, 23, 32, 39, 41, 42
satellite, 45
scarcity, xx
sea-level, xviii, 33, 51, 53, 55, 61
seasonality, 6, 42
sediment, xvii, 1, 3, 6, 10, 12, 14, 16, 21, 24, 27, 30, 32, 34, 43, 47, 49, 60, 63, 65
sediments, xv, xvii, 2, 3, 12, 14, 15, 16, 21, 27, 33, 43, 48, 49, 54, 56, 57
sensitivity, xvii
sewage, 50
shape, 32
signals, 1
silica, 27
similarity, 36
species, xv, xvii, xviii, xix, 1, 3, 7, 8, 9, 11, 12, 18, 20, 21, 27, 28, 33, 40, 43, 47, 53, 55, 58, 59
speed, 10
stability, 60
stabilization, 60
statistical inference, xix
strategies, 42
stratification, 60
stress, 47
striatum, 31
summer, 18, 20
supply, xvii
surface area, 10
susceptibility, 6
swamps, 33

## T

taphonomy, 57, 58
taxonomy, 8
temperature, xvii, 12, 22, 47
territory, 5
tides, 3, 35
training, 38
transgression, 13
transition, 1
transport, 2, 3, 7, 49

turnover, xix

## U

ulna, 29, 92, 93
urban centers, 18
urban settlement, 16

## V

values, 8, 9, 17, 21, 22, 23, 28, 38, 39, 41, 42, 60
variability, 22, 27, 32, 42, 57
variables, xvii, xix, 8, 9, 20, 22, 40, 49, 50, 59
variation, 19, 28, 60
vegetation, 43, 63, 65

## W

water quality, 8, 12, 14, 17, 48
wetlands, xviii, 58, 60
wind, 32
winter, 18, 20, 57